MATH

Grade 6

Thomas J. Richards
Mathematics Teacher
Lamar Junior-Senior High School
Lamar, Missouri

This book is dedicated to our children — Alyx, Nathan, Fred S., Dawn, Molly, Ellen, Rashaun, Brianna, Michele, Bradley, BriAnne, Kristie, Caroline, Dominic, Corey, Lindsey, Spencer, Morgan, Brooke, Cody, Sydney — and to all children who deserve a good education and who love to learn.

McGraw-Hill Consumer Products

**McGraw-Hill
Consumer Products**
A Division of The McGraw·Hill Companies

Copyright © 1997 McGraw-Hill Consumer Products.
Published by McGraw-Hill Learning Materials, an imprint of
McGraw-Hill Consumer Products.

Printed in the United States of America. All rights reserved.
Except as permitted under the United States Copyright Act, no
part of this publication may be reproduced or distributed in any
form or by any means, or stored in a database retrieval
system, without prior written permission from the publisher.

Send all inquiries to:
McGraw-Hill Consumer Products
250 Old Wilson Bridge Road
Worthington OH 43085

ISBN 1-57768-116-9

5 6 7 8 9 10 POH 03 02 01 00 99

Table of Contents

Scope and Sequence Charts .... iv–v
Using This Book vi
Facts Tests—Addition .... vii–viii
Facts Tests—Subtraction .... ix–x
Facts Tests—Multiplication .... xi–xii
Facts Tests—Division .... xiii–xiv
Assignment Record Sheet xv
Scoring Chart for Tests xvi

Chapter 1
Addition and Subtraction
(of whole numbers)
Pre-Test 1–2
Lessons 1–5 3–10
Chapter 1 Test 11

Chapter 2
Multiplication and Division
(of whole numbers)
Pre-Test 12
Lessons 1–7 13–24
Chapter 2 Test 25

Chapter 3
Multiplication
(of fractions)
Pre-Test 26
Lessons 1–11 27–40
Chapter 3 Test 41

Chapter 4
Addition and Subtraction
(of fractions)
Pre-Test 42
Lessons 1–7 43–52
Chapter 4 Test 53

Chapter 5
Division
(of fractions)
Pre-Test 54
Lessons 1–6 55–64
Chapter 5 Test 65

Chapter 6
Addition and Subtraction
(of decimals)
Pre-Test 66
Lessons 1–10 67–80
Chapter 6 Test 81

Chapter 7
Multiplication
(of decimals)
Pre-Test 82
Lessons 1–6 83–90
Chapter 7 Test 91

Chapter 8
Division
(of decimals)
Pre-Test 92
Lessons 1–7 93–104
Chapter 8 Test 105

Chapter 9
Metric Measurement
Pre-Test 106
Lessons 1–8 107–116
Chapter 9 Test 117

Chapter 10
Customary Measurement
Pre-Test 118
Lessons 1–5 119–126
Chapter 10 Test 127

Chapter 11
Percent
Pre-Test 128
Lessons 1–5 129–136
Chapter 11 Test 137

Chapter 12
Geometry
Pre-Test 138
Lessons 1–3 139–141
Chapter 12 Test 142

Test—Chapters 1–6 143–144
Final Test—Chapters 1–12 145–148
Answers to Lessons 149–156
Answers to Pre-Tests and Tests 157–160

iii

The SPECTRUM
Contents

GRADE ONE

Numeration
(0 through 10) 1–10

Addition and Subtraction
(Facts through 5)
11–18

Addition and Subtraction
(Facts through 8)
19–30

Addition and Subtraction
(Facts through 10)
31–44

Numeration
(through 99) 45–56

Time, Calendar,
Centimeter,
Inch 57–70

Addition and Subtraction
(2 digit; no renaming)
71–88

Addition and Subtraction
(Facts through 18)
89–110

Mid-Book/Final
Checkups 111–116

Answers 117–123

GRADE TWO

Addition and Subtraction
(Facts through 10)
1–9

Numeration
(through 99) 10–19

Addition and Subtraction
(Facts through 18)
20–35

Fractions, Time,
Calendar,
Centimeter, Inch
36–51

Addition and Subtraction
(2 digit; no renaming)
52–71

Addition and Subtraction
(2 digit; with renaming)
72–89

Numeration, Addition
and Subtraction
(3 digit; no renaming)
90–110

Mid-Book/Final
Checkups 111–116

Answers 117–123

GRADE THREE

Basic Facts Tests
vii–xiv

Addition and Subtraction
1–17

Addition and Subtraction
(2 digit; no renaming)
18–29

Addition and Subtraction
(2 digit; renaming)
30–41

Addition and Subtraction
(2–3 digit; renaming)
42–57

Addition and Subtraction
(3–4 digit; renaming)
58–67

Calendar, Time, Roman
Numerals, Money
68–75

Multiplication
(through 5 × 9) 76–85

Multiplication
(through 9 × 9) 86–95

Multiplication
(2 digit by 1 digit)
96–105

Division
(through 45 ÷ 5)
106–115

Division
(through 81 ÷ 9)
116–125

Metric Measurement
(linear, liquid)
126–133

Measurement
(linear, liquid, time)
134–144

Mid-Book/Final Tests
145–150

Answers 151–159

GRADE FOUR

Basic Facts Tests
vii–xiv

Addition and Subtraction
(1 and 2 digit; no
renaming) 1–11

Addition and Subtraction
(2 and 3 digit;
renaming) 12–25

Addition and Subtraction
(3 digit through 5 digit)
26–35

Multiplication
(3 digit by 1 digit)
36–47

Multiplication
(2 digit by 2 digit
through 3 digit by 2
digit) 48–59

Multiplication
(4 digit by 1 digit; 4
digit by 2 digit; 3 digit
by 3 digit) 60–71

Temperature and Money
72–79

Division
(basic facts) 80–89

Division
(2 and 3 digit
dividends) 90–103

Division
(4 digit dividends)
104–109

Multiplication and
Division 110–119

Metric Measurement
120–131

Measurement
(linear, capacity,
weight, time) 132–144

Mid-Book/Final Tests
145–150

Answers 151–160

MATHEMATICS Series of Units

GRADE FIVE

Basic Facts Tests vii–xiv

Addition and Subtraction (2 digit through 6 digit) 1–13

Multiplication (2 digit by 1 digit through 4 digit by 3 digit) 14–25

Division (2, 3, and 4 digit dividends) 26–35

Division (2, 3, and 4 digit dividends) 36–47

Division (4 and 5 digit dividends) 48–57

Metric Measurement 58–69

Measurement (linear, area, capacity, weight) 70–77

Fractional Numbers (fractions, mixed numerals, simplest form) 78–87

Multiplication (fractional numbers) 88–101

Addition (fractional numbers) 102–119

Subtraction (fractional numbers) 120–135

Geometry 136–142

Mid-Book/Final Tests 143–148

Answers 149–157

GRADE SIX

Basic Facts Tests vii–xiv

Addition and Subtraction (whole numbers) 1–11

Multiplication and Division (whole numbers) 12–25

Multiplication (fractions) 26–41

Addition and Subtraction (fractions) 42–53

Division (fractions) 54–65

Addition and Subtraction (decimals) 66–81

Multiplication (decimals) 82–91

Division (decimals) 92–105

Metric Measurement 106–117

Measurement (linear, capacity, weight, time) 118–127

Percent 128–137

Geometry 138–144

Mid-Book/Final Test 142–148

Answers 149–159

GRADE SEVEN

Problem-Solving Strategies (optional) vii–xiv

Addition, Subtraction, Multiplication, and Division (whole numbers) 1–25

Addition, Subtraction, Multiplication, and Division (fractional numbers) 26–49

Addition, Subtraction, Multiplication, and Division (decimals) 50–71

Ratio and Proportion 72–83

Decimals, Fractions, Percent 84–95

Percent 96–107

Interest (simple) 108–117

Metric Measurement 118–125

Geometry 126–135

Perimeter and Area (rectangles, triangles, circles) 136–151

Volume (rectangular solids, triangular prisms, cylinders) 152–159

Statistics and Probability 160–174

Mid-Book/Final Tests 175–180

Answers 181–191

GRADE EIGHT

Problem-Solving Strategies (optional) vii–xiv

Addition, Subtraction, Multiplication, and Division 1–23

Equations 24–37

Using Equations to Solve Problems 38–47

Ratio, Proportion, Percent 48–67

Simple and Compound Interest 68–77

Metric Measurement 78–87

Measurement and Approximation 88–103

Geometry 104–113

Similar Triangles and the Pythagorean Theorem 114–129

Perimeter, Area, and Volume 130–143

Graphs 144–155

Probability 156–172

Mid-Book/Final Tests 173–178

Answers 179–191

Using This Book

SPECTRUM MATHEMATICS is a non-graded, consumable series for students who need special help with the basic skills of computation and problem solving. This successful series emphasizes skill development and practice, without complex terminology or abstract symbolism. Because of the nature of the content and the students for whom the series is intended, readability has been carefully controlled to comply with the mathematics level of each book.

Features:
- A **Pre-Test** at the beginning of each chapter helps determine a student's understanding of the chapter content. The Pre-Test enables students and teachers to identify specific skills that need attention.
- **Developmental exercises** are provided at the top of the page when new skills are introduced. These exercises involve students in learning and serve as an aid for individualized instruction or independent study.
- **Abundant opportunities for practice** follow the developmental exercises.
- **Problem-solving pages** enable students to apply skills to realistic problems they will meet in everyday life.
- A **Test** at the end of each chapter gives students and teachers an opportunity to check understanding. A **Mid-Book Test**, covering Chapters 1–6, and a **Final Test**, covering all chapters, provide for further checks of understanding.
- A **Record of Test Scores** is provided on page xvi of this book so students can chart their progress as they complete each chapter test.
- **Answers** to all problems and test items are included at the back of the book.

This is the third edition of *SPECTRUM MATHEMATICS*. The basic books have remained the same. Some new, useful features have been added.

New Features:
- **Scope and Sequence Charts** for the entire *Spectrum Mathematics* series are included on pages iv–v.
- **Basic Facts Tests** for addition, subtraction, multiplication, and division are included on pages vii–xiv. There are two forms of each test. These may be given at any time the student or teacher decides they are appropriate.
- An **Assignment Record Sheet** is provided on page xv.

NAME _____

Addition Facts (Form A)

	a	b	c	d	e	f	g	h
1.	8 +2	3 +2	6 +2	2 +8	9 +2	4 +2	1 +2	7 +1
2.	5 +4	7 +9	1 +1	6 +9	4 +3	5 +8	5 +1	2 +0
3.	2 +7	0 +0	5 +2	3 +3	9 +1	1 +4	7 +8	6 +0
4.	6 +3	3 +1	7 +7	0 +1	2 +2	6 +8	5 +7	2 +9
5.	4 +4	8 +8	1 +6	8 +3	2 +6	8 +9	0 +3	7 +6
6.	9 +3	2 +5	7 +5	4 +9	6 +7	3 +4	8 +7	3 +9
7.	5 +3	3 +8	5 +9	3 +5	9 +7	5 +6	4 +8	9 +8
8.	1 +8	9 +5	0 +4	8 +4	3 +6	7 +4	6 +6	2 +3
9.	6 +4	4 +5	7 +3	6 +5	9 +6	0 +8	4 +6	8 +6
10.	8 +5	3 +7	2 +4	9 +4	4 +7	7 +2	9 +9	5 +5

Perfect score: 80   My score: _____

NAME _____

Addition Facts (Form B)

	a	b	c	d	e	f	g	h
1.	7 +2	5 +0	1 +5	8 +2	3 +2	6 +2	4 +9	2 +1
2.	8 +3	2 +9	9 +1	6 +3	4 +2	7 +0	3 +0	9 +4
3.	1 +7	7 +3	4 +8	2 +8	8 +1	1 +1	9 +9	3 +9
4.	8 +9	3 +3	0 +0	6 +9	2 +2	8 +0	3 +8	7 +7
5.	4 +3	6 +4	8 +4	5 +2	9 +2	4 +7	6 +8	1 +0
6.	7 +4	2 +3	1 +9	8 +8	0 +2	7 +6	5 +9	9 +3
7.	3 +4	9 +6	5 +3	7 +5	4 +6	8 +7	2 +7	3 +7
8.	5 +4	6 +5	0 +6	2 +4	8 +5	3 +6	7 +8	5 +8
9.	9 +8	3 +5	7 +9	4 +4	6 +6	5 +5	0 +9	2 +5
10.	5 +7	8 +6	2 +6	9 +7	5 +6	9 +5	4 +5	6 +7

Perfect score: 80   My score: _____

NAME _____

Subtraction Facts (Form A)

	a	b	c	d	e	f	g	h
1.	11 −8	9 −5	10 −7	7 −6	8 −4	15 −9	9 −8	13 −7
2.	13 −4	9 −9	15 −6	7 −4	4 −2	17 −8	8 −2	10 −6
3.	14 −9	7 −2	10 −5	3 −2	2 −1	10 −8	6 −4	13 −9
4.	16 −8	6 −2	11 −3	7 −3	9 −7	11 −4	8 −5	11 −6
5.	14 −7	4 −1	12 −4	5 −4	10 −2	15 −7	8 −3	14 −8
6.	12 −3	9 −4	11 −5	7 −7	7 −5	14 −6	8 −0	12 −5
7.	11 −7	6 −0	17 −9	5 −2	0 −0	13 −8	9 −3	10 −4
8.	12 −9	4 −0	15 −8	6 −3	10 −1	18 −9	6 −1	10 −3
9.	13 −6	5 −3	13 −5	2 −0	11 −2	12 −7	9 −6	11 −9
10.	16 −7	8 −6	12 −8	8 −1	16 −9	12 −6	9 −2	14 −5

Perfect score: 80   My score: _____

Subtraction Facts (Form B)

NAME _____

	a	b	c	d	e	f	g	h
1.	10 −7	4 −2	7 −3	15 −6	3 −1	8 −7	10 −5	11 −8
2.	10 −8	8 −3	9 −0	14 −9	6 −4	6 −3	11 −7	16 −7
3.	13 −9	5 −2	9 −5	11 −6	7 −0	8 −5	11 −5	15 −9
4.	12 −6	5 −1	5 −5	13 −4	5 −3	7 −4	15 −8	13 −7
5.	17 −8	9 −7	4 −3	14 −5	9 −2	9 −6	12 −8	10 −6
6.	12 −7	12 −9	8 −4	11 −4	6 −2	9 −3	16 −9	12 −4
7.	14 −6	10 −2	1 −0	16 −8	8 −6	7 −5	15 −7	10 −9
8.	14 −7	10 −4	3 −3	17 −9	7 −1	2 −1	13 −6	13 −5
9.	14 −8	11 −3	7 −2	12 −5	0 −0	6 −5	12 −3	11 −2
10.	18 −9	8 −8	9 −1	13 −8	8 −2	9 −4	11 −9	10 −3

Perfect score: 80   My score: _____

NAME _____

Multiplication Facts (Form A)

	a	b	c	d	e	f	g	h
1.	7 ×2	4 ×8	0 ×4	9 ×1	5 ×3	3 ×3	2 ×7	4 ×1
2.	8 ×6	1 ×5	5 ×2	3 ×2	7 ×0	6 ×6	7 ×9	1 ×3
3.	1 ×7	6 ×5	0 ×6	4 ×7	2 ×6	0 ×2	5 ×4	9 ×0
4.	8 ×7	2 ×8	7 ×8	9 ×2	4 ×2	5 ×5	8 ×5	3 ×9
5.	6 ×4	0 ×0	9 ×3	3 ×4	8 ×4	6 ×7	2 ×5	7 ×6
6.	7 ×3	9 ×4	4 ×6	6 ×3	6 ×8	4 ×9	7 ×7	1 ×1
7.	2 ×9	3 ×5	8 ×3	5 ×6	2 ×1	4 ×3	9 ×6	7 ×5
8.	9 ×9	4 ×5	6 ×9	0 ×8	9 ×8	6 ×2	8 ×2	2 ×4
9.	3 ×6	2 ×2	5 ×7	8 ×8	8 ×1	9 ×5	3 ×8	6 ×1
10.	5 ×8	7 ×4	4 ×4	3 ×7	2 ×3	9 ×7	5 ×9	8 ×9

Perfect score: 80   My score: _____

Multiplication Facts (Form B)

NAME _____

	a	b	c	d	e	f	g	h
1.	6 ×2	2 ×2	4 ×5	7 ×8	5 ×7	8 ×8	3 ×2	2 ×0
2.	9 ×2	5 ×6	3 ×3	6 ×3	1 ×9	4 ×4	7 ×7	1 ×4
3.	4 ×6	8 ×9	0 ×9	2 ×3	7 ×6	3 ×1	5 ×8	9 ×6
4.	1 ×0	3 ×4	7 ×5	6 ×4	5 ×5	8 ×7	6 ×5	4 ×3
5.	7 ×9	4 ×7	2 ×9	1 ×2	9 ×5	2 ×4	7 ×4	6 ×0
6.	3 ×5	8 ×6	4 ×8	7 ×3	5 ×4	4 ×2	0 ×7	9 ×7
7.	7 ×2	1 ×6	9 ×4	3 ×6	8 ×0	4 ×9	8 ×5	2 ×5
8.	2 ×8	5 ×3	5 ×9	4 ×0	9 ×8	2 ×6	3 ×7	1 ×8
9.	6 ×8	8 ×2	6 ×7	9 ×9	3 ×8	8 ×4	6 ×6	5 ×1
10.	3 ×9	9 ×3	0 ×5	2 ×7	5 ×2	7 ×1	8 ×3	6 ×9

Perfect score: 80   My score: _____

Division Facts (Form A)

	a	b	c	d	e	f	g
1.	4)16	1)6	8)16	2)10	3)18	4)36	4)4
2.	1)1	6)54	1)7	5)45	9)36	5)35	9)27
3.	8)8	4)12	3)15	7)0	8)24	2)12	4)20
4.	2)8	5)40	9)45	6)48	9)18	5)30	3)0
5.	3)27	1)8	7)63	1)5	4)0	7)14	8)32
6.	9)54	4)32	9)9	6)0	2)14	6)42	8)40
7.	7)28	2)6	5)25	7)21	7)56	2)2	5)5
8.	9)0	4)8	9)63	6)36	8)48	6)12	1)0
9.	5)20	3)3	7)35	2)16	4)28	3)12	7)49
10.	2)4	6)30	8)72	3)21	9)72	6)18	8)56
11.	4)24	1)4	5)15	1)2	7)42	1)3	3)9
12.	1)9	3)24	2)18	6)24	8)64	5)10	9)81

Perfect score: 84   My score: _____

Division Facts (Form B)

	a	b	c	d	e	f	g
1.	3)6	5)35	7)21	1)5	8)0	2)6	8)72
2.	6)24	8)8	3)9	9)54	5)30	1)4	6)42
3.	5)40	7)63	6)36	2)8	4)20	8)64	3)3
4.	2)10	4)24	4)4	9)0	1)6	5)45	8)16
5.	9)45	3)21	8)56	1)7	3)12	9)63	2)2
6.	6)30	5)25	2)0	7)56	2)4	7)14	4)16
7.	5)0	9)36	6)18	3)24	6)0	3)15	7)49
8.	1)2	8)24	2)12	8)48	9)72	4)12	1)3
9.	9)27	4)28	7)42	4)8	5)15	1)8	9)9
10.	1)1	5)20	3)27	6)48	7)28	6)12	8)40
11.	7)35	2)14	9)81	1)9	4)36	5)10	2)18
12.	4)32	6)6	8)32	3)18	9)18	2)16	6)54

Perfect score: 84   My score: _____

Assignment Record Sheet

NAME _____

Pages Assigned	Date	Score	Pages Assigned	Date	Score	Pages Assigned	Date	Score

SPECTRUM MATHEMATICS

Record of Test Scores

Rank	Test Pages													
	11	25	41	53	65	81	91	105	117	127	137	142	143–4	145–8
Excellent	25	25	20	20	20	25	25	20	20	25	25	20	50	100
Very Good	20	20	15	15	15	20	20	15	15	20	20	15	40	80
Good	15	15	10	10	10	15	15	10	10	15	15	10	30	60
Fair	10	10	5	5	5	10	10	5	5	10	10	5	20	40
Poor	5	5	0	0	0	5	5	0	0	5	5	0	10	20

To record the score you receive on a TEST:

(1) Find the vertical scale below the page number of that TEST,
(2) on that vertical scale, draw a ● at the mark which represents your score.

For example, if your score for the TEST on page 11 is "My score: 15," draw a ● at the 15-mark on the first vertical scale. A score of 15 would show that your rank is "Good." You can check your progress from one test to the next by connecting the dots with a line segment.

PRE-TEST—Addition and Subtraction

NAME _____

Chapter 1

Add or subtract.

	a	b	c	d	e	f
1.	35 +3	7 +43	43 +25	67 +28	73 +52	59 +63
2.	46 −5	57 −9	28 −13	148 −63	175 −86	214 −35
3.	421 +348	325 +436	783 +192	752 +638	428 +173	976 +544
4.	738 −125	872 −439	986 −394	1465 −938	1831 −256	3814 −915

Add or subtract.

	a	b	c	d	e
5.	4218 +3570	5831 +4179	6281 +3982	7543 +9647	2796 +8215
6.	7832 −4701	4216 −2437	52614 −8316	38126 −9433	42713 −5816
7.	53246 +32512	42186 +17287	38743 +45382	20917 +34216	52843 +28379
8.	82165 −31042	32186 −9178	42514 −3495	88672 −32967	98135 −28459
9.	42 26 +38	523 416 +758	4281 3826 +1435	42163 5286 +25488	32815 12916 +38442

Perfect score: 49   My score: _____

Pre-Test Problem Solving

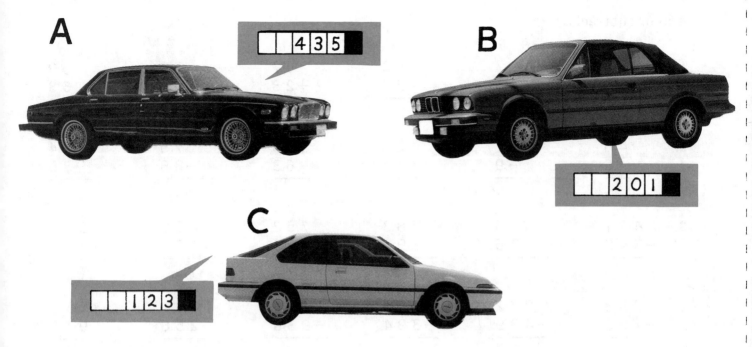

Solve each problem.

1. Odometer readings, such as shown above, tell how many miles a car has been driven. What is the total number of miles cars A and B have been driven?

Car A has been driven _____ miles.

Car B has been driven _____ miles.

Both cars have been driven _____ miles.

2. How many more miles has car A been driven than car C?

Car A has been driven _____ miles.

Car C has been driven _____ miles.

Car A has been driven _____ more miles.

3. What is the total number of miles car A, car B, and car C have been driven?

They have been driven _____ miles.

Perfect score: 7   My score: _____

Lesson 1 Addition

```
    5 7
      4
   +6 8
   ─────
```

Add the ones.
Rename 19 as 10 + 9.

```
    7            5̇ 7
    4              4
   +8           +6 8
   ───          ─────
   1 9 or 10 + 9   9
```

Add the tens.

```
   ⁱ5 7
      4
   +6 8
   ─────
   1 2 9
```

Add.

	a	b	c	d	e	f
1.	3 4 +5	6 +2 1	4 8 +5	9 +3 6	5 6 +3	9 +7 8
2.	3 5 +2 4	4 6 +3 2	3 7 +4 1	2 0 +5 8	3 1 +3 8	6 5 +1 3
3.	5 7 +2 4	3 6 +2 7	5 8 +1 9	5 2 +9 4	3 2 +7 1	5 5 +9 3
4.	3 5 +8 9	7 2 +7 9	8 6 +4 5	4 8 +6 3	3 7 +6 5	5 4 +9 8
5.	3 5 3 +2 1	2 7 1 8 +3 5	4 2 3 +7 0	5 2 1 6 +5 9	3 5 2 7 +6	5 8 3 7 +2 9
6.	3 6 8 4 2 7 +3 9	4 2 5 9 2 6 +7	2 1 8 5 4 +2 6	8 3 7 5 8 +7 5	3 1 8 0 6 0 +9	5 4 5 4 5 4 +5 4

Perfect score: 36   My score: _____

Lesson 2 Subtraction

NAME _____

Rename 72 as "6 tens and 12 ones."
Subtract the ones.

Rename 1 hundred and 6 tens as "16 tens."
Subtract the tens.

```
   1 7 2          1 ⁶7̶ ¹²2̶         ¹⁶1̶ ⁶7̶ ¹²2̶
  − 9 6          −  9 6          −   9 6
  ─────          ─────────        ─────────
                       6             7 6
```

Subtract.

	a	b	c	d	e	f
1.	57 −3	98 −4	63 −3	75 −6	38 −9	46 −8
2.	68 −25	75 −32	98 −44	37 −12	46 −36	58 −27
3.	53 −28	84 −36	61 −27	37 −18	25 −18	42 −25
4.	154 −27	193 −37	295 −27	146 −39	253 −27	104 −45
5.	163 −93	253 −62	357 −71	176 −83	483 −93	519 −34
6.	184 −97	352 −69	463 −87	108 −29	520 −83	645 −96

Perfect score: 36   My score: _____

Lesson 3 Addition

Add the ones.	Add the tens.	Add the hundreds. Rename 16 hundreds as "1 thousand and 6 hundreds."	Add the thousands.

3 3 ¹6
   6 2 8
+8 7 3 7
─────────
           1
	3 3 ¹6
	   6 2 8
	+8 7 3 7
	─────────
	         8 1
		¹3 3 ¹6
		   6 2 8
		+8 7 3 7
		─────────
		       6 8 1
			¹3 3 ¹6
			   6 2 8
			+8 7 3 7
			─────────
			1 2 6 8 1

Add.

	a	b	c	d	e
1.	423 +165	527 +319	382 +476	528 +739	524 +898
2.	3168 +3240	3782 +4561	8093 +1279	5837 +2896	6789 +4567
3.	54312 +24241	52168 +29210	83245 +13876	42104 +49863	54372 +36798
4.	423 104 +735	1423 3410 +6578	4216 3807 +4218	42116 38425 +10731	22430 38654 +12465
5.	52316 4284 +3721	342 1563 +78216	1423 7386 +214	42305 316 +4217	5203 48310 +244
6.	342 153 786 +403	2173 4168 5246 +3700	4215 3800 2407 +3142	12421 33568 45423 +13154	12421 13685 17256 +60381
7.	421 3863 425 +38160	1725 42311 3820 +421	7526 380 62776 +5381	4270 38197 244 +42311	738 52865 376 +25400

Perfect score: 35   My score: _____

Problem Solving

Solve each problem.

1. A trucker drove 528 kilometers on the first trip and 746 kilometers on the next. How many kilometers did the trucker drive altogether?

_____ kilometers were driven on the first trip.

_____ kilometers were driven on the next trip.

_____ kilometers were driven altogether.

2. Xemo Corporation filled 5,281 orders last week and 7,390 orders this week. How many orders were filled in these two weeks?

_____ orders were filled last week.

_____ orders were filled this week.

_____ orders were filled in the two weeks.

3. Xemo Corporation produced 42,165 xemos in January and 34,895 xemos in February. How many xemos were produced in both January and February?

_____ xemos were produced in January.

_____ xemos were produced in February.

_____ xemos were produced in both months.

4. In three weeks Mr. Jenkins carried the following number of passengers on his bus: 4,216; 3,845; and 7,281. What was the total number of passengers carried in the three weeks?

_____ passengers were carried.

5. The odometer readings on the last three cars that Mr. Williams sold were 22,163; 48,395; and 23,842. How many miles were recorded on these three cars?

_____ miles were recorded.

6. Last month three jets were flown the following number of miles: 42,816; 5,421; and 38,652. What was the total number of miles flown?

_____ miles were flown.

Perfect score: 12   My score: _____

Lesson 4 Subtraction

NAME _____

	Subtract the ones.	Rename. Subtract the tens.	Subtract the hundreds.	Continue to rename and subtract as needed.
	42017 −3846 ——— 1	4 2 0 1 7 −3 8 4 6 ——— 7 1	4 2 0 1 7 −3 8 4 6 ——— 1 7 1	4 2 0 1 7 −3 8 4 6 ——— 3 8 1 7 1

Subtract.

	a	b	c	d	e
1.	736 −324	546 −329	831 −480	516 −337	703 −299
2.	4216 −314	2468 −539	5468 −573	2345 −456	1306 −457
3.	5246 −2215	3872 −2438	4351 −2263	4020 −1706	7503 −2455
4.	53211 −4298	42683 −3167	54216 −5299	60831 −7081	29540 −5219
5.	42465 −21528	38429 −14953	76543 −37835	82106 −47297	30907 −18608
6.	67230 −41195	42007 −18246	86992 −20997	71549 −10856	90036 −89595

Perfect score: 30   My score: _____

Problem Solving

Solve each problem.

1. It takes 500 points to win a prize. Pat has 385 points now. How many more points does she need to win a prize?

   _____ points are needed to win a prize.

   Pat now has _____ points.

   She needs _____ more points.

2. There are 1,516 pupils enrolled at Webb School. Of these, 842 are girls. How many are boys?

   _____ pupils are enrolled.

   _____ of the pupils are girls.

   _____ of the pupils are boys.

3. Factory A employs 5,281 people and factory B employs 3,817 people. How many more people does factory A employ than factory B?

   Factory A employs _____ more people.

4. Mr. Wells had 52,816 miles on his car when he traded it. The car he traded for has 4,357 miles on it. How many fewer miles does it have than the older car?

   It has _____ fewer miles.

5. Last year 42,169 orders were shipped from a warehouse. So far this year 5,837 orders have been shipped. How many more orders must be shipped this year in order to match the total for last year?

   _____ more orders must be shipped.

6. The odometer on Jim's car reads 52,116. The odometer on Pat's car reads 38,429. How many more miles are on Jim's car than are on Pat's car?

   _____ more miles are on Jim's car.

Perfect score: 10   My score: _____

Lesson 5 Addition and Subtraction

Add or subtract.

	a	b	c	d	e	f
1.	32 +6	5 +48	23 +35	47 +26	89 +50	78 +57
2.	58 −3	72 −21	47 −38	159 −93	143 −85	202 −37
3.	523 +364	428 +537	683 +194	385 +276	483 +629	753 +869
4.	783 −502	926 −418	564 −283	1925 −137	2436 −648	1926 −928

Add or subtract.

	a	b	c	d	e
5.	5231 +3468	4661 +2179	3157 +6930	2087 +9237	4281 +6759
6.	8426 −3312	7531 −3452	8426 −2756	13041 −9158	25308 −8499
7.	63125 +10420	42163 +45387	28135 +47385	61702 +28715	37839 +57893
8.	72519 −30418	83162 −35087	52083 −41839	98035 −68746	63613 −55895
9.	23 34 +42	426 709 +358	4216 5384 +2196	22514 43868 +21706	82965 372 +1451

Perfect score: 49   My score: _____

Problem Solving

Answer each question.

1. In a contest, Mary earned 758 points. Helen earned 929 points. Bill earned 1,356 points. How many points did the two girls earn?

Are you to add or subtract? _____

How many points did the two girls earn? _____

2. In problem 1, how many more points did Bill earn than Helen?

Are you to add or subtract? _____
How many more points did
  Bill earn than Helen? _____

3. In problem 1, how many points did all three people earn?

Are you to add or subtract? _____

How many points did all three earn? _____

4. This month 32,526 people visited the museum. Last month 28,831 people visited the museum. How many more people visited the museum this month than last month?

Are you to add or subtract? _____
How many more people visited the
  museum this month than last month? _____

5. In problem 4, how many people visited the museum during the two months?

Are you to add or subtract? _____
How many people visited the
  museum during the two months? _____

6. At the beginning of last year 52,116 cars were registered. There were 4,913 new cars registered the first six months and 3,085 the second six months. How many cars were registered at the end of the year?

Are you to add or subtract? _____
How many cars were registered
  at the end of the year? _____

1.

2.

3.

4.

5.

6.

Perfect score: 12   My score: _____

10

CHAPTER 1 TEST

Add or subtract.

	a	b	c	d	e
1.	42 +9	75 +83	96 +58	147 +129	345 +286
2.	54 −6	39 −27	158 −79	384 −215	580 −483
3.	4216 −3817	15382 −8293	42165 −38479	52163 −44318	84362 −53977
4.	5421 +8892	5843 +6969	52816 +32558	4235 6815 +42916	38433 12758 +28906

Answer each question.

5. At the end of last year, suppose the odometer reading on your car was 33,384. You drove your car 29,458 kilometers last year. What was the reading at the beginning of last year?

Are you to add or subtract? _____
What was the reading at
   the beginning of last year? _____

6. In problem 5, suppose you expect to drive the car the same number of kilometers this year as you did last year. If you do, what will the reading be at the end of this year?

Are you to add or subtract? _____
What will the reading be
   at the end of this year? _____

7. Suppose the distances you drove in the last three years were 42,516 kilometers, 38,342 kilometers, and 14,208 kilometers. How many kilometers did you drive in three years?

How many kilometers did
   you drive in three years? _____

Perfect score: 25   My score: _____

PRE-TEST—Multiplication and Division

NAME _____

Chapter 2

Multiply.

	a	b	c	d	e
1.	33 ×3	48 ×4	304 ×2	432 ×8	1234 ×2
2.	6789 ×5	133 ×21	456 ×34	1231 ×22	5783 ×45
3.	123 ×321	576 ×435	1302 ×132	4563 ×478	5009 ×837

Divide.

4. 25)225 14)518 27)463 14)4550 95)3610

5. 53)7832 92)12420 58)45530 32)78216 73)52914

Perfect score: 25   My score: _____

Lesson 1 Multiplication

Multiply.

	a	b	c	d	e	f	g	h
1.	4 ×0	2 ×0	8 ×0	1 ×0	7 ×1	6 ×1	1 ×1	5 ×1
2.	8 ×2	2 ×2	4 ×2	7 ×2	6 ×2	5 ×2	3 ×2	9 ×2
3.	9 ×3	7 ×3	5 ×3	0 ×3	1 ×3	6 ×3	4 ×3	3 ×3
4.	4 ×4	3 ×4	5 ×4	8 ×4	7 ×4	0 ×4	9 ×4	1 ×4
5.	8 ×5	2 ×5	7 ×5	5 ×5	4 ×5	3 ×5	1 ×5	0 ×5
6.	8 ×6	2 ×6	9 ×6	7 ×6	6 ×6	5 ×6	1 ×6	3 ×6
7.	9 ×7	7 ×7	6 ×7	0 ×7	1 ×7	5 ×7	8 ×7	4 ×7
8.	0 ×8	5 ×8	8 ×8	9 ×8	4 ×8	3 ×8	6 ×8	7 ×8
9.	3 ×9	9 ×9	8 ×9	1 ×9	2 ×9	7 ×9	6 ×9	4 ×9

Perfect score: 72   My score: _____

Lesson 2 Division

Divide.

	a	b	c	d	e	f	g	h
1.	1)2	1)3	1)5	1)4	1)6	1)9	1)8	1)1
2.	2)18	2)12	2)14	2)16	2)8	2)10	2)4	2)2
3.	3)0	3)15	3)9	3)12	3)24	3)18	3)3	3)21
4.	4)20	4)8	4)4	4)12	4)32	4)24	4)36	4)16
5.	5)30	5)45	5)0	5)10	5)25	5)15	5)40	5)5
6.	6)30	6)24	6)42	6)6	6)12	6)36	6)54	6)48
7.	7)0	7)21	7)14	7)56	7)49	7)63	7)35	7)28
8.	8)16	8)0	8)56	8)72	8)48	8)32	8)24	8)40
9.	9)45	9)27	9)36	9)63	9)9	9)81	9)0	9)54

Perfect score: 72   My score: _____

Lesson 3   Multiplication

Multiply 7 ones by 5.

```
   3
9 8 1 7̸     7
    ×5     ×5
    ———    ———
     5      35
```

Multiply 1 ten by 5. Add the 3 tens.

```
   3
9 8 1̸ 7    10
    ×5     ×5
    ———    ———
   8 5     50
          +30
          ———
           80
```

Multiply 8 hundreds by 5.

```
 4 3
9 8̸ 1 7    800
    ×5     ×5
    ———    ————
 0 8 5     4000
```

Multiply 9 thousands by 5. Add the 4 thousands.

```
 4 3
9̸ 8 1 7    9000
    ×5     ×5
    ————   ————
49 0 8 5   45000
          +4000
          —————
           49000
```

Multiply.

	a	b	c	d	e
1.	32 ×3	23 ×4	82 ×3	78 ×8	95 ×6
2.	421 ×2	123 ×4	241 ×3	501 ×5	159 ×6
3.	783 ×3	538 ×8	762 ×5	954 ×7	473 ×9
4.	1033 ×2	3216 ×3	3172 ×3	5014 ×2	3257 ×3
5.	1478 ×6	5738 ×7	4826 ×9	5384 ×6	7083 ×5

Perfect score: 25   My score: _____

Problem Solving

Solve each problem.

1. Mrs. Clarke has 24 employees. Each employee makes 5 units each day. How many units do all employees complete in one day?

There are _____ employees.

Each employee makes _____ units each day.

The employees make _____ units each day.

2. Each bus can carry 77 passengers. How many passengers can be carried on 7 such buses?

Each bus can carry _____ passengers.

There are _____ buses in all.

A total of _____ passengers can be carried.

3. There are 365 days in a year, except leap year which has 366 days. How many days are there in 3 years if there is no leap year included?

There are _____ days in a year.

The number of days in _____ years is to be found.

There are _____ days in 3 years.

4. Seven hundred seventy-five meals were prepared each day for 5 days. How many meals were prepared in the 5 days?

_____ meals were prepared in the 5 days.

5. Michael earned 3,401 points. His sister earned twice as many. How many points did his sister earn?

His sister earned _____ points.

6. A machine is designed to produce 2,965 parts each day. How many parts should the machine produce in 7 days?

The machine should produce _____ parts in 7 days.

1.	
2.	
3.	4.
5.	6.

Perfect score: 12   My score: _____

16

Lesson 4 Multiplication

Multiply 4567 by 1.

```
  4567
×  321
  4567
```

Multiply 4567 by 20.

```
  4567
×  321
  4567
 91340
```

Multiply 4567 by 300.

```
   4567
×   321
   4567
  91340
1370100
```

```
   4567
×   321
   4567  ⎫
  91340  ⎬ Add.
1370100  ⎭
1466007
```

Multiply.

	a	b	c	d	e
1.	57 ×21	48 ×32	75 ×63	135 ×48	276 ×42
2.	531 ×27	835 ×92	1864 ×27	3186 ×54	7083 ×92

Multiply.

	a	b	c	d
3.	413 ×214	564 ×532	217 ×416	908 ×592
4.	1564 ×795	3827 ×630	9216 ×205	5043 ×684

Perfect score: 18   My score: _____

Problem Solving

Solve each problem.

1. Each box weighs 28 kilograms. What is the weight of 35 such boxes?

Each box weighs _____ kilograms.

There are _____ boxes in all.

The total weight is _____ kilograms.

2. There are 19 carpenters working for a construction firm. Each worked 47 hours last week. What is the total number of hours they worked last week?

Each carpenter worked _____ hours.

There are _____ carpenters in all.

_____ hours were worked.

3. The production schedule estimates that 321 machines can be produced each week. At that rate, how many machines can be produced in 52 weeks?

There are _____ machines scheduled to be produced each week.

There are _____ weeks.

_____ machines can be produced in 52 weeks.

4. The rail distance between Los Angeles and New York is 3,257 miles. How many miles would a train travel if it made 32 one-way trips between these two cities?

The train would travel _____ miles.

5. There are 731 cases of zoopers in the warehouse. Each case contains 144 zoopers. How many zoopers are in the warehouse?

There are _____ zoopers in the warehouse.

6. There are 1,440 minutes in one day. How many minutes are in 365 days?

There are _____ minutes in 365 days.

Perfect score: 12   My score: _____

Lesson 5   Division

NAME _____

Study how to divide 2074 by 6.

×	100	200	300	400
6	600	1200	1800	2400

2074 is between 1800 and 2400, so 2074 ÷ 6 is between 300 and 400. The hundreds digit is 3.

```
      3
   ┌─────
  6│ 2074
    1800      (300 × 6)
    ────
     274      Subtract.
```

×	10	20	30	40	50
6	60	120	180	240	300

274 is between 240 and 300, so 274 ÷ 6 is between 40 and 50. The tens digit is 4.

```
      34
   ┌─────
  6│ 2074
    1800
    ────
     274
     240      (40 × 6)
     ────
      34      Subtract.
```

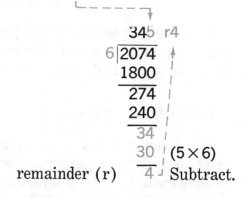

×	1	2	3	4	5	6	7
6	6	12	18	24	30	36	42

34 is between 30 and 36, so 34 ÷ 6 is between 5 and 6. The ones digit is 5.

```
      345  r4
   ┌─────
  6│ 2074
    1800
    ────
     274
     240
     ────
      34
      30     (5 × 6)
      ──
       4     Subtract.
```

remainder (r)

Divide.

	a	b	c	d	e
1.	4)92	3)58	3)72	4)77	6)810
2.	3)225	6)590	6)8080	9)4739	6)4254

Perfect score: 10   My score: _____

Problem Solving

Solve each problem.

1. There are 5 people at each table. Two people are standing. There are 92 people in the room. How many tables are there?

   There are _____ people in the room.

   There are _____ people at each table.

   There are _____ tables in the room.

2. Three people earned 774 points in a contest. Suppose each person earned the same number of points. How many points did each person earn?

   Each person earned _____ points.

3. As each new car comes off an assembly line, it receives 8 gallons of gasoline. How many new cars can receive gasoline from a tank containing 2,440 gallons?

   _____ new cars can receive gasoline.

4. Four baseballs are put in each box. How many boxes are needed to package 273 baseballs? How many baseballs would be left?

   _____ boxes are needed.

   _____ baseball would be left.

5. A train travels 6,516 miles to make a round trip between New York and Los Angeles. How many miles would the train travel from Los Angeles to New York?

   The train would travel _____ miles.

6. Each carton holds 8 bottles. How many full cartons could be filled with 3,075 bottles? How many bottles would be left over?

   _____ cartons could be filled.

   _____ bottles would be left over.

Perfect score: 10   My score: _____

Lesson 6 Division

Study how to divide 28888 by 95.

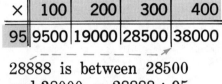

28888 is between 28500 and 38000, so 28888 ÷ 95 is between 300 and 400. The hundreds digit is 3.

```
      3
95|28888
   28500   (300×95)
     388   Subtract.
```

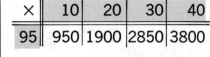

Since 388 is less than 950, the tens digit is 0.

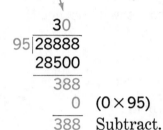

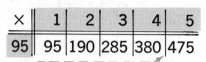

388 is between 380 and 475, so 388 ÷ 95 is between 4 and 5. The ones digit is 4.

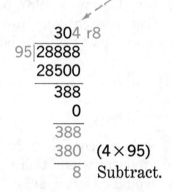

Divide.

	a	b	c	d	e
1.	25)810	33)891	18)819	27)727	75)6900
2.	54)7695	28)9698	98)34937	75)39400	42)14742

Perfect score: 10   My score: _____

21

Problem Solving

Solve each problem.

1. There are 988 units to be shipped. Each crate will hold 26 units. How many crates will be needed to ship all the units?

There are _____ units to be shipped.

Each crate will hold _____ units.

_____ crates will be needed.

2. Mr. Lodey has 987 parts to pack. He will pack 24 parts in each box. How many boxes will he need? How many parts will be left over?

He will need _____ boxes.

He will have _____ parts left over.

3. A bank considers 30 days to be a month. How many months would there be in 9,295 days? How many days would be left over?

There would be _____ months.

There would be _____ days left over.

4. During a two-week period, 75 employees worked a total of 5,625 hours. Each employee worked the same number of hours. How many hours did each employee work?

Each employee worked _____ hours.

5. There are 76 sections with a total of 17,100 seats in the new stadium. Each section has the same number of seats. How many seats are in each section?

There are _____ seats in each section.

6. Three dozen grapefruit are packed in a case. How many cases would be needed to pack 27,100 grapefruit? How many grapefruit would be left over?

_____ cases would be needed.

_____ grapefruit would be left over.

Perfect score: 11   My score: _____

Lesson 7 Multiplication and Division

Multiply.

	a	b	c	d	e
1.	35 ×7	347 ×5	385 ×8	1438 ×9	4906 ×6
2.	37 ×85	48 ×54	357 ×92	289 ×38	4356 ×27
3.	158 ×132	706 ×315	345 ×221	709 ×743	689 ×838

Divide.

4. 22)386   35)452   25)950   12)1468   54)8478

5. 15)3092   75)4728   37)15725   53)24815   43)45683

Perfect score: 25   My score: _____

Problem Solving

Answer each question.

1. It takes 75 hours to make one tractor. How many hours would it take to make 650 tractors?

   Are you to multiply or divide? _____
   How many hours would
     it take to make 650 tractors? _____

2. It takes 28 minutes to make one hubcap. How many hubcaps could be made in 196 minutes?

   Are you to multiply or divide? _____
   How many hubcaps could
     be made in 196 minutes? _____

3. There are 168 hours in one week. How many hours are in 260 weeks?

   Are you to multiply or divide? _____
   How many hours are
     in 260 weeks? _____

4. There were 5,790 tickets sold at the game. There are 75 tickets on a roll of tickets. How many complete rolls of tickets were sold? How many tickets from the next roll were sold?

   Are you to multiply or divide? _____
   How many complete
     rolls of tickets were sold? _____
   How many tickets from the
     next roll were sold? _____

5. A satellite orbits the moon every 45 minutes. How many complete orbits could it make in 5,545 minutes? How many minutes would be left over?

   Are you to multiply or divide? _____
   How many complete
     orbits could be made? _____
   How many minutes
     would be left over? _____

6. There are 10,080 minutes in a week. How many minutes are in 52 weeks?

   Are you to multiply or divide? _____
   How many minutes
     are in 52 weeks? _____

1.	2.
3.	4.
5.	6.

Perfect score: 14   My score: _____

CHAPTER 2 TEST

Multiply.

	a	b	c	d	e
1.	43 ×2	38 ×9	507 ×8	1351 ×6	7254 ×7
2.	35 ×23	48 ×76	155 ×33	2056 ×42	3718 ×37
3.	304 ×144	215 ×255	1403 ×304	5555 ×246	3182 ×354

Divide.

4. 34)136   12)420   53)781   37)4316   29)1702

5. 74)9990   46)1257   38)38640   14)53821   63)27342

Perfect score: 25   My score: _____

PRE-TEST—Multiplication

Chapter 3

Write each answer in simplest form.

	a	b	c	d
1.	$\frac{1}{2} \times \frac{3}{5}$	$\frac{4}{7} \times \frac{4}{5}$	$\frac{2}{3} \times \frac{5}{7}$	$\frac{2}{5} \times \frac{2}{5}$
2.	$\frac{3}{5} \times \frac{1}{6}$	$\frac{3}{8} \times \frac{5}{9}$	$\frac{6}{7} \times \frac{3}{8}$	$\frac{8}{9} \times \frac{3}{10}$
3.	$3 \times \frac{2}{5}$	$5 \times \frac{8}{9}$	$6 \times \frac{3}{4}$	$\frac{8}{9} \times 3$
4.	$4 \times 3\frac{1}{3}$	$2\frac{1}{2} \times 5$	$4 \times \frac{1}{6}$	$\frac{7}{8} \times 12$
5.	$2\frac{1}{3} \times 1\frac{1}{4}$	$1\frac{7}{8} \times 1\frac{2}{7}$	$4\frac{2}{3} \times 1\frac{3}{7}$	$3\frac{1}{3} \times 2\frac{2}{5}$

Perfect score: 20   My score: _____

Lesson 1 Fractions

NAME _____

     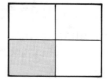

$\frac{2}{3}$ ← **numerator** ← − − − −number of parts colored − − − − → numerator → ___
     ← **denominator** ← − number of parts of the same size − → denominator →

$\frac{2}{3}$ of the triangle is colored.   ____ of the square is colored.

Write the fraction that tells how much of each figure is colored.

          a   b   c   d

1. ___   ___   ___   ___

2. ___

Draw a line segment between each fraction and number word that name the same number.

        a   b

3. one half $\frac{4}{5}$ three eighths $\frac{7}{9}$

4. two thirds $\frac{3}{4}$ four sevenths $\frac{3}{8}$

5. three fourths $\frac{2}{3}$ three sevenths $\frac{3}{7}$

6. four fifths $\frac{1}{2}$ seven eighths $\frac{4}{7}$

7. five sixths $\frac{5}{6}$ seven ninths $\frac{7}{8}$

Write a fraction for each of the following.

        a   b

8. numerator 4, denominator 7 _____   three fifths _____

9. numerator 5, denominator 8 _____   two sevenths _____

10. denominator 10, numerator 9 _____   four ninths _____

Perfect score: 24   My score: _____

Lesson 2 Mixed Numerals

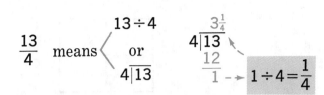

$3\frac{1}{4}$ is a short way to write $3+\frac{1}{4}$.

$3\frac{1}{4}$ is a **mixed numeral**.

Complete the following.

	a	b	c	d
1.	$3\frac{1}{5} = 3 + \underline{}$	$4\frac{1}{2} = \underline{} + \frac{1}{2}$	$3\frac{3}{4} = \underline{} + \underline{}$	$9 + \frac{1}{3} = \underline{}$
2.	$4\frac{2}{3} = 4 + \underline{}$	$5\frac{3}{7} = \underline{} + \frac{3}{7}$	$6\frac{2}{5} = \underline{} + \underline{}$	$8 + \frac{7}{8} = \underline{}$
3.	$5\frac{1}{8} = 5 + \underline{}$	$2\frac{1}{6} = \underline{} + \frac{1}{6}$	$3\frac{1}{3} = \underline{} + \underline{}$	$5 + \frac{3}{7} = \underline{}$

Change each fraction to a mixed numeral.

	a	b	c
4.	$\frac{5}{2}$	$\frac{9}{5}$	$\frac{7}{2}$
5.	$\frac{9}{4}$	$\frac{6}{5}$	$\frac{8}{3}$
6.	$\frac{14}{3}$	$\frac{10}{3}$	$\frac{17}{5}$

Tell whether each of the following is *less than 1, equal to 1,* or *greater than 1*.

	a	b	c
7.	$\frac{7}{8}$	$\frac{5}{4}$	$\frac{6}{6}$
8.	$\frac{2}{3}$	$\frac{12}{12}$	$\frac{11}{10}$
9.	$\frac{1}{9}$	$\frac{12}{9}$	$\frac{10}{5}$

Perfect score: 33   My score: _____

Lesson 3   Addition

$\frac{2}{5}+\frac{1}{5}=\frac{2+1}{5}$ Add the numerators.
$\phantom{\frac{2}{5}+\frac{1}{5}}=\frac{3}{5}$ Use the same denominator.

$+\frac{2}{5}$
$+\frac{1}{5}$
$\overline{\frac{3}{5}}$

$\frac{3}{10}+\frac{4}{10}+\frac{2}{10}=\frac{++}{10}$
$\phantom{\frac{3}{10}+\frac{4}{10}+\frac{2}{10}}=\frac{}{10}$

$\frac{3}{10}$
$\frac{4}{10}$
$+\frac{2}{10}$
$\overline{\phantom{+\frac{2}{10}}}$

Add.

	a	b	c	d
1.	$\frac{3}{5}+\frac{1}{5}$	$\frac{4}{8}+\frac{3}{8}$	$\frac{2}{7}+\frac{2}{7}$	$\frac{1}{5}+\frac{2}{5}+\frac{1}{5}$
2.	$\frac{3}{6}+\frac{2}{6}$	$\frac{1}{7}+\frac{3}{7}$	$\frac{2}{8}+\frac{1}{8}$	$\frac{1}{4}+\frac{1}{4}+\frac{1}{4}$
3.	$\frac{3}{10}+\frac{4}{10}$	$\frac{4}{12}+\frac{1}{12}$	$\frac{5}{11}+\frac{4}{11}$	$\frac{2}{15}+\frac{2}{15}+\frac{7}{15}$

Add.

	a	b	c	d	e	f
4.	$\frac{4}{6}$ $+\frac{1}{6}$	$\frac{3}{8}$ $+\frac{4}{8}$	$\frac{1}{7}$ $+\frac{2}{7}$	$\frac{3}{10}$ $+\frac{6}{10}$	$\frac{7}{12}$ $+\frac{4}{12}$	$\frac{3}{11}$ $+\frac{1}{11}$
5.	$\frac{1}{5}$ $\frac{1}{5}$ $+\frac{1}{5}$	$\frac{2}{7}$ $\frac{3}{7}$ $+\frac{1}{7}$	$\frac{2}{8}$ $\frac{1}{8}$ $+\frac{2}{8}$	$\frac{4}{10}$ $\frac{1}{10}$ $+\frac{2}{10}$	$\frac{3}{15}$ $\frac{4}{15}$ $+\frac{4}{15}$	$\frac{1}{12}$ $\frac{4}{12}$ $+\frac{2}{12}$

Perfect score: 24   My score: _____

Lesson 4 Changing a Mixed Numeral to a Fraction

$4\frac{2}{3} = \frac{(3 \times 4) + 2}{3}$ Multiply the denominator by the whole number and add the numerator. $3\frac{1}{6} = \frac{(\times) + }{6}$

$= \frac{12 + 2}{3}$ Use the same denominator. $= \frac{ + }{6}$

$= \frac{14}{3}$ $= \frac{}{6}$

Change each mixed numeral to a fraction.

	a	b	c
1.	$2\frac{5}{8}$	$2\frac{3}{5}$	$3\frac{2}{3}$
2.	$3\frac{7}{10}$	$10\frac{2}{3}$	$14\frac{1}{2}$
3.	$6\frac{7}{8}$	$5\frac{9}{10}$	$13\frac{5}{12}$
4.	$4\frac{5}{6}$	$7\frac{3}{4}$	$8\frac{11}{12}$

Perfect score: 12   My score: _____

Lesson 5 Multiplication

Multiply the numerators.

$$\frac{2}{3} \times \frac{1}{5} = \frac{2 \times 1}{3 \times 5} = \frac{2}{15}$$

Multiply the denominators.

$$\frac{1}{2} \times \frac{3}{4} = \frac{1 \times 3}{2 \times 4} \qquad \qquad \frac{2}{5} \times \frac{1}{3} = \frac{\times}{\times}$$

$$= \frac{}{8} \qquad \qquad = \frac{}{}$$

Multiply.

	a	b	c	d
1.	$\frac{1}{2} \times \frac{1}{3}$	$\frac{3}{4} \times \frac{1}{2}$	$\frac{1}{3} \times \frac{1}{4}$	$\frac{3}{5} \times \frac{1}{2}$
2.	$\frac{3}{5} \times \frac{3}{4}$	$\frac{4}{7} \times \frac{3}{5}$	$\frac{4}{5} \times \frac{2}{3}$	$\frac{3}{8} \times \frac{5}{7}$
3.	$\frac{2}{3} \times \frac{4}{5}$	$\frac{1}{8} \times \frac{1}{2}$	$\frac{5}{7} \times \frac{3}{4}$	$\frac{3}{5} \times \frac{7}{8}$
4.	$\frac{6}{7} \times \frac{3}{5}$	$\frac{2}{9} \times \frac{1}{3}$	$\frac{5}{8} \times \frac{3}{7}$	$\frac{2}{7} \times \frac{3}{5}$
5.	$\frac{7}{8} \times \frac{7}{8}$	$\frac{2}{3} \times \frac{2}{3}$	$\frac{4}{9} \times \frac{2}{3}$	$\frac{4}{5} \times \frac{6}{7}$
6.	$\frac{8}{9} \times \frac{5}{7}$	$\frac{5}{8} \times \frac{1}{3}$	$\frac{5}{6} \times \frac{5}{7}$	$\frac{3}{8} \times \frac{5}{8}$

Perfect score: 24   My score: _____

Lesson 6 Renaming Numbers

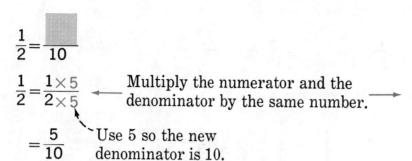

$\dfrac{1}{2} = \dfrac{\square}{10}$

$\dfrac{1}{2} = \dfrac{1 \times 5}{2 \times 5}$ ← Multiply the numerator and the denominator by the same number. →

$= \dfrac{5}{10}$ Use 5 so the new denominator is 10.

$4 = \dfrac{\square}{3}$

$4 = \dfrac{4}{1}$ Name the whole number as a fraction.

$= \dfrac{4 \times 3}{1 \times 3}$

$= \dfrac{12}{3}$ Use 3 so the new denominator is 3.

Rename as shown.

       a   b   c   d

1. $\dfrac{1}{2} = \dfrac{4}{8}$ $\dfrac{2}{3} = \dfrac{\square}{6}$ $\dfrac{4}{5} = \dfrac{\square}{10}$ $\dfrac{3}{4} = \dfrac{\square}{12}$

$\dfrac{1}{2} = \dfrac{1 \times 4}{2 \times 4}$

$= \dfrac{4}{8}$

2. $3 = \dfrac{12}{4}$ $4 = \dfrac{\square}{6}$ $3 = \dfrac{\square}{10}$ $5 = \dfrac{\square}{5}$

$\dfrac{3}{1} = \dfrac{3 \times 4}{1 \times 4}$

$= \dfrac{12}{4}$

3. $\dfrac{1}{2} = \dfrac{\square}{16}$ $10 = \dfrac{\square}{8}$ $\dfrac{5}{8} = \dfrac{\square}{16}$ $\dfrac{5}{6} = \dfrac{\square}{18}$

Perfect score: 10   My score: _____

Lesson 7 Greatest Common Factor

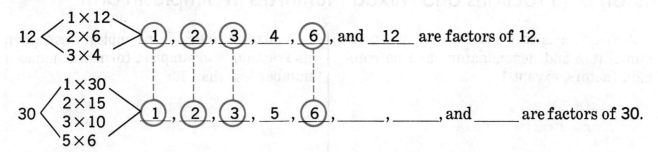

1, 2, 3, and 6 are **common factors** of 12 and 30.

6 is the **greatest common factor** of 12 and 30.

List the factors of each number named below. Then list the common factors and the greatest common factor of each pair of numbers.

		factors	common factor(s)	greatest common factor
1.	6	_____		
	10	_____	_____	_____
2.	5	_____		
	8	_____	_____	_____
3.	12	_____		
	15	_____	_____	_____
4.	10	_____		
	20	_____	_____	_____
5.	14	_____		
	16	_____	_____	_____
6.	15	_____		
	7	_____	_____	_____
7.	24	_____		
	18	_____	_____	_____

Perfect score: 28   My score: _____

Lesson 8 Fractions and Mixed Numerals in Simplest Form

A fraction is in simplest form when its numerator and denominator have no common factors, except 1.

Divide 12 and 15 by their greatest common factor.

$$\frac{12}{15} = \frac{12 \div 3}{15 \div 3} = \frac{4}{5}$$

The simplest form for $\frac{12}{15}$ is $\frac{4}{5}$.

A mixed numeral is in simplest form when its fraction is in simplest form and names a number less than 1.

Divide 4 and 6 by their greatest common factor.

$$3\frac{4}{6} = 3 + \frac{4 \div 2}{6 \div 2}$$
$$= 3 + \frac{2}{3}$$
$$= 3\frac{2}{3}$$

The simplest form for $3\frac{4}{6}$ is _____.

Change each of the following to simplest form.

	a	b	c
1.	$\frac{8}{10}$	$\frac{10}{20}$	$\frac{14}{21}$
2.	$2\frac{4}{8}$	$3\frac{6}{9}$	$5\frac{8}{10}$
3.	$\frac{12}{18}$	$5\frac{9}{12}$	$\frac{15}{18}$
4.	$6\frac{8}{12}$	$\frac{25}{30}$	$3\frac{12}{16}$
5.	$\frac{24}{30}$	$3\frac{14}{18}$	$\frac{16}{32}$

Perfect score: 15   My score: _____

Lesson 9 Multiplication

$$\frac{1}{2} \times \frac{3}{4} = \frac{1 \times 3}{2 \times 4}$$
$$= \frac{3}{8}$$

Is $\frac{3}{8}$ in simplest form? _____

$$\frac{4}{5} \times \frac{1}{6} = \frac{4 \times 1}{5 \times 6}$$
$$= \frac{4}{30} \longrightarrow \frac{4}{30} = \frac{4 \div 2}{30 \div 2}$$
$$= \frac{2}{15} \longleftarrow = \frac{2}{15}$$

Is $\frac{4}{30}$ in simplest form? _____

Is $\frac{2}{15}$ in simplest form? _____

Write each answer in simplest form.

	a	b	c	d
1.	$\frac{1}{2} \times \frac{3}{5}$	$\frac{2}{3} \times \frac{4}{5}$	$\frac{2}{3} \times \frac{2}{3}$	$\frac{5}{6} \times \frac{1}{7}$
2.	$\frac{3}{4} \times \frac{4}{5}$	$\frac{5}{6} \times \frac{2}{3}$	$\frac{6}{7} \times \frac{2}{3}$	$\frac{3}{5} \times \frac{4}{9}$
3.	$\frac{5}{6} \times \frac{2}{5}$	$\frac{4}{5} \times \frac{5}{6}$	$\frac{3}{8} \times \frac{2}{3}$	$\frac{2}{10} \times \frac{5}{6}$
4.	$\frac{6}{5} \times \frac{3}{8}$	$\frac{9}{10} \times \frac{5}{12}$	$\frac{8}{9} \times \frac{3}{10}$	$\frac{5}{6} \times \frac{9}{10}$
5.	$\frac{4}{7} \times \frac{5}{6}$	$\frac{3}{8} \times \frac{7}{10}$	$\frac{9}{10} \times \frac{5}{9}$	$\frac{6}{7} \times \frac{9}{10}$

Perfect score: 20   My score: _____

Problem Solving

Solve. Write each answer in simplest form.

1. The Urbans had $\frac{3}{4}$ gallon of milk. One half of this was used for dinner. How much milk was used for dinner? ($\frac{1}{2}$ of $\frac{3}{4} = \frac{1}{2} \times \frac{3}{4}$)

   _____ gallon was used for dinner.

2. Marsha read $\frac{4}{5}$ of a book. Two thirds of that reading was done at school. How much of the book did she read at school?

   She read _____ of the book at school.

3. Tricia lives $\frac{4}{5}$ mile from work. One morning she ran $\frac{1}{2}$ of the distance to work. How far did Tricia run?

   Tricia ran _____ mile.

4. Three fourths of a room has been painted. Joe did $\frac{2}{3}$ of the painting. How much of the room did Joe paint?

   Joe painted _____ of the room.

5. A truck was carrying $\frac{3}{4}$ ton of sand. One third of the sand was put into barrels. How much sand was put into barrels?

   _____ ton of sand was put into barrels.

6. Carrie had a rope that was $\frac{2}{3}$ yard long. She used $\frac{1}{2}$ of it. How much rope did she use?

   _____ yard of rope was used.

7. One fourth of the people in the room have blue eyes. Two thirds of the blue-eyed people have blond hair. What part of the people in the room have blond hair and blue eyes?

   _____ have blond hair and blue eyes.

Perfect score: 7   My score: _____

NAME _____

Lesson 10 Multiplication

$$4 \times \frac{5}{6} = \frac{4}{1} \times \frac{5}{6}$$
$$= \frac{4 \times 5}{1 \times 6}$$
$$= \frac{20}{6}$$
$$= 3\frac{1}{3}$$

Rename whole numbers and mixed numerals as fractions.

Multiply the fractions.

Change to simplest form.

$$4\frac{2}{3} \times 5 = \frac{14}{3} \times \frac{5}{1}$$
$$= \frac{14 \times 5}{3 \times 1}$$
$$= \frac{70}{3}$$
$$= 23\frac{1}{3}$$

Write each answer in simplest form.

	a	b	c	d
1.	$5 \times \frac{2}{3}$	$6 \times \frac{4}{5}$	$\frac{1}{2} \times 9$	$\frac{3}{4} \times 7$
2.	$9 \times \frac{5}{6}$	$\frac{1}{4} \times 6$	$\frac{3}{8} \times 12$	$10 \times \frac{4}{5}$
3.	$2\frac{1}{2} \times 3$	$1\frac{1}{3} \times 5$	$2 \times 3\frac{2}{5}$	$4 \times 4\frac{2}{3}$

Perfect score: 12   My score: _____

Problem Solving

Solve. Write each answer in simplest form.

1. A can of fruit weighs $\frac{2}{3}$ pound. How many pounds would 3 cans of fruit weigh?

Three cans of fruit would weigh _____ pounds.

2. A plumber expects a job to take 10 hours. The plumber has already worked $\frac{4}{5}$ of that time. How many hours has the plumber worked?

The plumber has worked _____ hours.

3. Each book is $\frac{7}{8}$ inch thick. How many inches high would a stack of 12 such books be?

The stack would be _____ inches high.

4. The carpenter stacked 15 sheets of wall board on top of each other. If each sheet is $\frac{5}{8}$ inch thick, how high is the stack?

The stack is _____ inches high.

5. Mark practiced the piano for $\frac{3}{4}$ hour on each of 4 days. How many hours did he practice in all?

Mark practiced _____ hours in all.

6. Each hamburger weighs $\frac{1}{4}$ pound. How much will 6 hamburgers weigh?

Six hamburgers will weigh _____ pounds.

7. There are 24 people at a meeting. Suppose $\frac{2}{3}$ of the people are women. How many of the people are women? How many are men?

_____ of the people are women.

_____ of the people are men.

Perfect score: 8   My score: _____

Lesson 11 Multiplication

$2\frac{3}{5} \times 1\frac{1}{6} = \frac{13}{5} \times \frac{7}{6}$ Change the mixed numerals to fractions.

$\phantom{2\frac{3}{5} \times 1\frac{1}{6}} = \frac{13 \times 7}{5 \times 6}$ Multiply the fractions.

$\phantom{2\frac{3}{5} \times 1\frac{1}{6}} = \frac{91}{30}$

$\phantom{2\frac{3}{5} \times 1\frac{1}{6}} = 3\frac{1}{30}$ Change to simplest form.

Write each answer in simplest form.

	a	b	c	d
1.	$4\frac{2}{3} \times 1\frac{2}{5}$	$3\frac{1}{2} \times 1\frac{1}{6}$	$1\frac{2}{3} \times 2\frac{1}{2}$	$2\frac{2}{3} \times 2\frac{2}{3}$
2.	$2\frac{2}{5} \times 2\frac{1}{4}$	$1\frac{7}{10} \times 2\frac{1}{2}$	$5\frac{1}{3} \times 1\frac{1}{5}$	$2\frac{4}{5} \times 1\frac{1}{7}$
3.	$3\frac{3}{4} \times 2\frac{1}{3}$	$3\frac{2}{5} \times 1\frac{7}{8}$	$4\frac{2}{3} \times 1\frac{1}{8}$	$3\frac{3}{4} \times 3\frac{1}{3}$
4.	$5\frac{1}{6} \times 6\frac{3}{8}$	$2\frac{3}{5} \times 2\frac{1}{2}$	$1\frac{1}{4} \times 1\frac{1}{4}$	$3\frac{1}{8} \times 6\frac{2}{3}$

Perfect score: 16   My score: _____

Problem Solving

Solve. Write each answer in simplest form.

1. A full box of soap weighs $2\frac{2}{3}$ pounds. How many pounds would $1\frac{1}{3}$ boxes of soap weigh?

   They would weigh _____ pounds.

2. It takes $1\frac{4}{5}$ hours to process 1 ton of ore. How many hours would it take to process $3\frac{1}{3}$ tons of ore?

   It would take _____ hours.

3. Each box of bolts weighs $3\frac{3}{4}$ pounds. How many pounds would $8\frac{1}{2}$ boxes of bolts weigh?

   They would weigh _____ pounds.

4. The boys can walk $3\frac{1}{2}$ miles in 1 hour. At that rate, how many miles could the boys walk in $1\frac{1}{6}$ hours?

   The boys could walk _____ miles.

5. Each bag of apples weighs $4\frac{1}{2}$ pounds. How much would $3\frac{1}{2}$ bags of apples weigh?

   They would weigh _____ pounds.

6. Riding her bicycle, Terry averages $9\frac{1}{2}$ miles per hour. At that speed, how far could she go in $2\frac{2}{3}$ hours?

   She could go _____ miles.

7. In problem 6, suppose Terry averages $9\frac{3}{4}$ miles per hour. How far could she go in $2\frac{2}{3}$ hours?

   She could go _____ miles.

8. A machine can process $2\frac{1}{2}$ tons in 1 hour. How many tons can the machine process in $2\frac{1}{10}$ hours?

   The machine can process _____ tons in $2\frac{1}{10}$ hours.

9. If the machine in problem 8 broke down after $1\frac{1}{2}$ hours, how many tons would have been processed?

   _____ tons would have been processed.

1.	
2.	
3.	
4.	
5.	
6.	
7.	
8.	
9.	

Perfect score: 9   My score: _____

CHAPTER 3 TEST

NAME _____

Write each answer in simplest form.

	a	b	c	d
1.	$\frac{1}{2} \times \frac{5}{6}$	$\frac{7}{8} \times \frac{5}{6}$	$\frac{2}{3} \times \frac{5}{7}$	$\frac{3}{8} \times \frac{3}{8}$
2.	$\frac{5}{9} \times \frac{6}{7}$	$\frac{7}{10} \times \frac{8}{9}$	$\frac{9}{10} \times \frac{5}{6}$	$\frac{5}{8} \times \frac{4}{5}$
3.	$2 \times \frac{3}{5}$	$6 \times \frac{5}{7}$	$\frac{1}{2} \times 8$	$\frac{5}{6} \times 8$
4.	$4 \times 3\frac{1}{3}$	$\frac{4}{5} \times 2$	$10 \times \frac{4}{5}$	$\frac{3}{8} \times 10$
5.	$3\frac{1}{3} \times 1\frac{1}{7}$	$1\frac{4}{5} \times 3\frac{1}{2}$	$2\frac{2}{3} \times 1\frac{1}{10}$	$2\frac{4}{5} \times 4\frac{1}{6}$

Perfect score: 20   My score: _____

PRE-TEST—Addition and Subtraction NAME _____ Chapter 4

Write each answer in simplest form.

	a	b	c	d
1.	$\dfrac{3}{7} + \dfrac{1}{7}$	$\dfrac{4}{9} + \dfrac{2}{9}$	$\dfrac{7}{8} - \dfrac{2}{8}$	$\dfrac{9}{10} - \dfrac{3}{10}$
2.	$\dfrac{2}{3} + \dfrac{1}{2}$	$\dfrac{4}{6} + \dfrac{7}{12}$	$\dfrac{5}{6} - \dfrac{3}{8}$	$\dfrac{9}{10} - \dfrac{5}{8}$
3.	$3 - \dfrac{2}{5}$	$1 - \dfrac{7}{8}$	$4 + \dfrac{3}{5} + \dfrac{5}{6}$	$\dfrac{1}{2} + 3 + \dfrac{2}{5}$
4.	$2\dfrac{3}{4} + \dfrac{1}{2}$	$\dfrac{7}{10} + 3\dfrac{7}{8}$	$5\dfrac{4}{9} - \dfrac{1}{3}$	$2\dfrac{7}{12} - \dfrac{5}{6}$
5.	$5\dfrac{4}{9} - 3\dfrac{1}{6}$	$7\dfrac{2}{5} - 2\dfrac{9}{10}$	$2\dfrac{1}{2} + 4\dfrac{1}{3} + 3\dfrac{2}{5}$	$4\dfrac{1}{2} + \dfrac{5}{6} + 3\dfrac{2}{3}$

Perfect score: 20   My score: _____

Lesson 1   Addition and Subtraction

Add the numerators

$$\frac{7}{8}+\frac{5}{8}=\frac{7+5}{8}=\frac{12}{8}=1\frac{1}{2}$$

Use the same denominator.

$$\frac{7}{8}+\frac{5}{8}=\frac{12}{8}=1\frac{1}{2}$$

Subtract the numerators.

$$\frac{5}{6}-\frac{1}{6}=\frac{5-1}{6}=\frac{4}{6}=\frac{2}{3}$$

Use the same denominator.

$$\frac{5}{6}-\frac{1}{6}=\frac{4}{6}=\frac{2}{3}$$

Change to simplest form.

Write each answer in simplest form.

	a	b	c	d	e
1.	$\frac{1}{5}+\frac{2}{5}$	$\frac{4}{7}+\frac{2}{7}$	$\frac{3}{4}+\frac{2}{4}$	$\frac{5}{6}+\frac{4}{6}$	$\frac{7}{8}+\frac{7}{8}$
2.	$\frac{5}{6}-\frac{4}{6}$	$\frac{7}{8}-\frac{3}{8}$	$\frac{5}{7}-\frac{2}{7}$	$\frac{9}{9}-\frac{4}{9}$	$\frac{5}{8}-\frac{1}{8}$
3.	$\frac{3}{10}+\frac{6}{10}$	$\frac{8}{9}+\frac{4}{9}$	$\frac{3}{8}+\frac{3}{8}$	$\frac{5}{12}+\frac{5}{12}$	$\frac{10}{15}+\frac{14}{15}$
4.	$\frac{11}{12}-\frac{3}{12}$	$\frac{7}{8}-\frac{2}{8}$	$\frac{8}{9}-\frac{5}{9}$	$\frac{9}{10}-\frac{4}{10}$	$\frac{9}{16}-\frac{3}{16}$
5.	$\frac{7}{12}+\frac{8}{12}$	$\frac{5}{9}-\frac{2}{9}$	$\frac{8}{15}+\frac{10}{15}$	$\frac{7}{10}-\frac{3}{10}$	$\frac{6}{14}+\frac{6}{14}$

Perfect score: 25   My score: _____

Problem Solving

Solve. Write each answer in simplest form.

1. Walter drank $\frac{1}{4}$ gallon of milk yesterday and $\frac{1}{4}$ gallon of milk today. What part of a gallon of milk did he drink during these two days?

   He drank _____ gallon of milk.

2. Trina and Trudy have painted $\frac{3}{4}$ of a room. Trudy painted $\frac{1}{4}$ of the room. How much of the room did Trina paint?

   Trina painted _____ of the room.

3. Tom measured two boards. He found that each was $\frac{3}{8}$ inch thick. What would be the total thickness of the boards if he glues them together?

   The total thickness would be _____ inch.

4. An unopened box of cereal weighed $\frac{15}{16}$ pound. Mother used $\frac{5}{16}$ pound of cereal from the box. How much cereal remains in the box?

   _____ pound remains in the box.

5. A television show has just begun and will last $\frac{5}{6}$ hour. After $\frac{4}{6}$ hour, what part of an hour remains of the television show?

   _____ hour remains of the television show.

6. Helen jogged $\frac{3}{4}$ hour before work. That same day she jogged $\frac{3}{4}$ hour after work. How long did she jog in all that day?

   She jogged _____ hours that day.

7. Jerry spent $\frac{5}{6}$ hour typing. He spent $\frac{1}{6}$ hour proofreading his typing. How long did he spend typing and proofreading in all?

   He spent _____ hour typing and proofreading.

8. In problem 7, how much longer did he spend typing than proofreading?

   He spent _____ hour more typing than proofreading.

1.	2.
3.	4.
5.	6.
7.	8.

Perfect score: 8   My score: _____

Lesson 2 Addition and Subtraction

$$\frac{2}{3} \begin{array}{c}\times 2 \\ \times 2\end{array} = \frac{4}{6}$$
$$+\frac{1}{2} \begin{array}{c}\times 3 \\ \times 3\end{array} = +\frac{3}{6}$$
$$\frac{7}{6} = 1\frac{1}{6}$$

The denominators are 3 and 2. Since $2 \times 3 = 6$, rename each fraction with a denominator of 6.

Add or subtract the fractions.

Write the answer in simplest form.

$$\frac{2}{3} \begin{array}{c}\times 2 \\ \times 2\end{array} = \frac{4}{6}$$
$$-\frac{1}{2} \begin{array}{c}\times 3 \\ \times 3\end{array} = -\frac{3}{6}$$
$$\frac{1}{6}$$

Write each answer in simplest form.

	a	*b*	*c*	*d*
1.	$\frac{3}{5}$ $+\frac{2}{3}$	$\frac{5}{6}$ $+\frac{1}{5}$	$\frac{1}{2}$ $+\frac{1}{3}$	$\frac{3}{10}$ $+\frac{1}{3}$
2.	$\frac{2}{3}$ $-\frac{1}{4}$	$\frac{5}{6}$ $-\frac{2}{5}$	$\frac{7}{8}$ $-\frac{2}{3}$	$\frac{3}{4}$ $-\frac{1}{3}$
3.	$\frac{7}{8}$ $+\frac{1}{3}$	$\frac{7}{8}$ $-\frac{1}{3}$	$\frac{2}{5}$ $+\frac{3}{4}$	$\frac{1}{2}$ $-\frac{1}{3}$
4.	$\frac{1}{3}$ $+\frac{3}{4}$	$\frac{3}{5}$ $-\frac{1}{3}$	$\frac{1}{2}$ $+\frac{4}{5}$	$\frac{3}{4}$ $-\frac{2}{3}$

Perfect score: 16   My score: _____

Lesson 3 Addition and Subtraction

$\dfrac{3}{4}$ $\times 3 \atop \times 3$ $\dfrac{9}{12}$ The denominators are 4 and 6. Since $3 \times 4 = 12$ and $2 \times 6 = 12$, rename each fraction with a denominator of 12.
$+\dfrac{1}{6}$ $\times 2 \atop \times 2$ $+\dfrac{2}{12}$
$\dfrac{11}{12}$ Add the fractions.

$\dfrac{9}{10}$ $\longrightarrow$ $\dfrac{9}{10}$ The denominators are 5 and 10. Since $2 \times 5 = 10$, rename only $\dfrac{2}{5}$ with a denominator of 10. Subtract the fractions.
$-\dfrac{2}{5}$ $\times 2 \atop \times 2$ $-\dfrac{4}{10}$
$\dfrac{5}{10} = \dfrac{1}{2}$ Change to simplest form.

Write each answer in simplest form.

	a	b	c	d
1.	$\dfrac{2}{5}$ $+\dfrac{1}{2}$	$\dfrac{2}{5}$ $+\dfrac{1}{3}$	$\dfrac{3}{4}$ $+\dfrac{1}{2}$	$\dfrac{7}{8}$ $+\dfrac{1}{4}$
2.	$\dfrac{2}{3}$ $-\dfrac{1}{4}$	$\dfrac{3}{5}$ $-\dfrac{1}{2}$	$\dfrac{2}{3}$ $-\dfrac{1}{6}$	$\dfrac{1}{2}$ $-\dfrac{3}{10}$
3.	$\dfrac{9}{10}$ $+\dfrac{1}{2}$	$\dfrac{5}{6}$ $-\dfrac{3}{4}$	$\dfrac{5}{6}$ $+\dfrac{1}{2}$	$\dfrac{7}{10}$ $-\dfrac{1}{5}$
4.	$\dfrac{2}{3}$ $+\dfrac{5}{6}$	$\dfrac{11}{12}$ $-\dfrac{1}{4}$	$\dfrac{5}{6}$ $+\dfrac{3}{10}$	$\dfrac{9}{10}$ $-\dfrac{1}{2}$

Perfect score: 16   My score: _____

Lesson 4 Addition

$3\frac{1}{2} \longrightarrow 3\frac{4}{8}$
$+1\frac{1}{8} \longrightarrow +1\frac{1}{8}$
$\phantom{+1\frac{1}{8} \longrightarrow} 4\frac{5}{8}$

Rename the numbers so the fractions have the same denominator.
Add the fractions.
Add the whole numbers.

$1\frac{1}{2} \longrightarrow 1\frac{6}{12}$
$3\frac{3}{4} \longrightarrow 3\frac{9}{12}$
$+\frac{2}{3} \longrightarrow +\frac{8}{12}$

Change to simplest form. $4\frac{23}{12} = 5\frac{11}{12}$

Write each answer in simplest form.

	a	b	c	d
1.	$3\frac{1}{4}$ $+2\frac{4}{5}$	$3\frac{1}{6}$ $+\frac{3}{4}$	$5\frac{1}{2}$ $+1\frac{5}{8}$	$3\frac{11}{12}$ $+\frac{5}{6}$
2.	$9\frac{7}{8}$ $+\frac{3}{4}$	$7\frac{2}{5}$ $+4\frac{3}{10}$	$\frac{3}{5}$ $+2\frac{5}{6}$	$\frac{9}{10}$ $+3\frac{5}{6}$
3.	$6\frac{2}{3}$ $1\frac{3}{4}$ $+\frac{1}{6}$	$2\frac{1}{5}$ $2\frac{1}{4}$ $+1\frac{1}{2}$	$3\frac{1}{3}$ $\frac{5}{6}$ $+3\frac{7}{12}$	$\frac{1}{2}$ $5\frac{1}{5}$ $+1\frac{3}{10}$
4.	$\frac{3}{5}$ $1\frac{2}{3}$ $+2\frac{1}{2}$	$3\frac{5}{8}$ $2\frac{1}{6}$ $+\frac{5}{12}$	$\frac{1}{4}$ $1\frac{1}{2}$ $+4\frac{7}{8}$	$2\frac{2}{3}$ $2\frac{1}{2}$ $+3\frac{2}{5}$

Perfect score: 16   My score: _____

Lesson 5 Subtraction

$3\frac{2}{3} \rightarrow 3\frac{4}{6}$
$-1\frac{1}{6} \rightarrow -1\frac{1}{6}$
$\overline{} \quad\quad \overline{2\frac{3}{6}} = 2\frac{1}{2}$

Rename the numbers so the fractions have the same denominator.
Subtract the fractions.
Subtract the whole numbers.
Change to simplest form.

$3 \rightarrow 2\frac{4}{4}$
$-\frac{1}{4} \rightarrow -\frac{1}{4}$
$\overline{} \quad\quad \overline{2\frac{3}{4}}$

$3 = 2+1$
$= 2+\frac{4}{4}$
$= 2\frac{4}{4}$

Write each answer in simplest form.

	a	b	c	d
1.	7 $-\frac{3}{4}$	4 $-\frac{1}{2}$	5 $-\frac{2}{3}$	8 $-\frac{1}{8}$
2.	$3\frac{4}{5}$ $-1\frac{1}{2}$	$5\frac{2}{3}$ $-3\frac{4}{9}$	$4\frac{5}{6}$ $-1\frac{1}{2}$	$5\frac{9}{10}$ $-3\frac{2}{5}$
3.	5 $-\frac{3}{5}$	$6\frac{3}{4}$ $-5\frac{1}{8}$	$2\frac{2}{3}$ $-1\frac{1}{2}$	10 $-2\frac{3}{10}$
4.	$10\frac{5}{6}$ $-7\frac{5}{12}$	8 $-\frac{5}{8}$	$9\frac{5}{6}$ $-2\frac{1}{3}$	6 $-\frac{9}{10}$

Perfect score: 16   My score: _____

Lesson 6 Subtraction

NAME _____

Rename the numbers so the fractions have the same denominator.

Rename $3\frac{3}{12}$ so you can subtract the fractions.

$3\frac{1}{4} \rightarrow 3\frac{3}{12} \rightarrow 2\frac{15}{12}$
$-1\frac{5}{6} \rightarrow -1\frac{10}{12} \rightarrow -1\frac{10}{12}$
$\phantom{-1\frac{5}{6} \rightarrow -1\frac{10}{12} \rightarrow }1\frac{5}{12}$

$3\frac{3}{12} = 2 + 1\frac{3}{12}$
$\phantom{3\frac{3}{12}} = 2 + \frac{15}{12}$
$\phantom{3\frac{3}{12}} = 2\frac{15}{12}$

Rename $4\frac{5}{10}$ so you can subtract the fractions.

$4\frac{1}{2} \rightarrow 4\frac{5}{10} \rightarrow 3\frac{15}{10}$
$-1\frac{3}{5} \rightarrow -1\frac{6}{10} \rightarrow -1\frac{6}{10}$
$\phantom{-1\frac{3}{5} \rightarrow -1\frac{6}{10} \rightarrow }2\frac{9}{10}$

$4\frac{5}{10} = 3 + 1\frac{5}{10}$
$\phantom{4\frac{5}{10}} = 3 + \frac{15}{10}$
$\phantom{4\frac{5}{10}} = 3\frac{15}{10}$

Write each answer in simplest form.

	a	b	c	d
1.	$2\frac{1}{2}$ $-1\frac{3}{4}$	$5\frac{1}{4}$ $-4\frac{1}{3}$	$7\frac{3}{4}$ $-3\frac{5}{6}$	$6\frac{2}{5}$ $-4\frac{1}{2}$
2.	$8\frac{3}{8}$ $-1\frac{7}{8}$	$5\frac{1}{4}$ $-4\frac{3}{4}$	$3\frac{7}{12}$ $-1\frac{3}{4}$	$4\frac{1}{6}$ $-2\frac{3}{8}$
3.	$9\frac{1}{2}$ $-5\frac{7}{10}$	$6\frac{1}{3}$ $-3\frac{5}{6}$	$11\frac{1}{6}$ $-4\frac{5}{12}$	$3\frac{3}{10}$ $-2\frac{1}{2}$
4.	$7\frac{3}{4}$ $-5\frac{4}{5}$	$12\frac{1}{4}$ $-6\frac{5}{8}$	$15\frac{1}{2}$ $-8\frac{2}{3}$	$10\frac{1}{6}$ $-3\frac{3}{4}$

Perfect score: 16   My score: _____

Problem Solving

Solve. Write each answer in simplest form.

1. A record has been playing for $\frac{1}{3}$ hour. The record still has $\frac{1}{12}$ hour to play. What is the total length of time the record can play?

   The record can play _____ hour.

2. It rained $\frac{3}{4}$ inch yesterday and $\frac{3}{10}$ inch today. How much more did it rain yesterday?

   It rained _____ inch more yesterday.

3. Jim spent $\frac{1}{2}$ hour doing his history homework and $\frac{3}{4}$ hour doing his science homework. How much time did he spend doing homework?

   He spent _____ hours doing homework.

4. Bob has a board that is $\frac{1}{8}$ inch too wide. The board is $\frac{3}{4}$ inch wide. What width board does Bob need?

   Bob needs a board _____ inch wide.

5. Shirley read $\frac{3}{5}$ hour in the morning and $\frac{1}{2}$ hour in the afternoon. How many hours did she read in the morning and afternoon?

   She read _____ hours.

6. In problem 5, how much longer did she read in the morning than in the afternoon?

   She read _____ hour longer in the morning.

7. John has two boxes. One weighs $\frac{3}{10}$ pound and the other weighs $\frac{7}{8}$ pound. What is the combined weight of both boxes?

   The combined weight is _____ pounds.

8. In problem 7, how much more does the heavier box weigh?

   The heavier box weighs _____ pound more.

Perfect score: 8   My score: _____

Lesson 7 Addition and Subtraction

Write each answer in simplest form.

	a	b	c	d
1.	$\dfrac{7}{8}$ $+\dfrac{5}{8}$	$\dfrac{9}{16}$ $-\dfrac{3}{16}$	$\dfrac{3}{5}$ $+\dfrac{2}{5}$	$\dfrac{11}{12}$ $-\dfrac{3}{12}$
2.	$\dfrac{1}{2}$ $+\dfrac{7}{8}$	$\dfrac{4}{5}$ $-\dfrac{2}{3}$	$\dfrac{4}{9}$ $+\dfrac{5}{6}$	$\dfrac{3}{4}$ $-\dfrac{5}{12}$
3.	9 $-\dfrac{5}{9}$	1 $-\dfrac{3}{10}$	4 $\dfrac{3}{8}$ $+\dfrac{3}{4}$	$\dfrac{7}{12}$ $\dfrac{2}{3}$ $+7$
4.	$\dfrac{2}{3}$ $+3\dfrac{4}{5}$	$7\dfrac{3}{8}$ $-\dfrac{2}{3}$	$6\dfrac{1}{6}$ $+\dfrac{5}{12}$	$1\dfrac{2}{5}$ $-\dfrac{7}{10}$
5.	$14\dfrac{3}{4}$ $-3\dfrac{11}{12}$	$16\dfrac{11}{12}$ $-2\dfrac{1}{6}$	$3\dfrac{2}{3}$ $2\dfrac{1}{5}$ $+4\dfrac{3}{8}$	$2\dfrac{5}{6}$ $3\dfrac{1}{5}$ $+2\dfrac{3}{10}$

Perfect score: 20   My score: _____

Problem Solving

Today's Workouts	
Kerri	$1\frac{1}{2}$ hours
Jennifer	$\frac{3}{4}$ hour
Ahmad	2 hours
Risa	$\frac{2}{3}$ hour

Solve. Write each answer in simplest form.

1. Who had a longer workout, Jennifer or Risa? How much longer?

_____ had a longer workout.

It was _____ hour longer.

2. How much longer was Ahmad's workout than Kerri's workout?

Ahmad's workout was _____ hour longer.

3. Risa finished her workout just as Kerri started his. How long did it take from the time Risa started until Kerri finished?

It took _____ hours.

4. Kerri and Jennifer started their workouts at the same time. When Jennifer finished her workout, how much longer did Kerri have to finish her workout?

She had _____ hour left to finish her workout.

5. What was the total workout time for all four people on the list?

The total workout time was _____ hours.

6. What is the difference between the longest workout time and the shortest workout time?

The difference is _____ hours.

Perfect score: 7   My score: _____

CHAPTER 4 TEST

NAME _____

Write each answer in simplest form.

	a	b	c	d
1.	$\dfrac{3}{8}$ $+\dfrac{4}{8}$	$\dfrac{7}{10}$ $-\dfrac{2}{10}$	$\dfrac{5}{12}$ $+\dfrac{9}{12}$	$\dfrac{8}{9}$ $-\dfrac{6}{9}$
2.	$\dfrac{5}{6}$ $+\dfrac{2}{9}$	$\dfrac{7}{8}$ $+\dfrac{1}{3}$	$\dfrac{9}{10}$ $-\dfrac{2}{5}$	$\dfrac{3}{4}$ $-\dfrac{2}{3}$
3.	6 $-\dfrac{1}{9}$	3 $-\dfrac{4}{7}$	$\dfrac{3}{8}$ $\dfrac{5}{6}$ $+4$	$\dfrac{7}{8}$ 2 $+\dfrac{3}{4}$
4.	$4\dfrac{5}{6}$ $+\dfrac{3}{5}$	$\dfrac{3}{8}$ $+2\dfrac{9}{10}$	$6\dfrac{3}{8}$ $-\dfrac{5}{6}$	$7\dfrac{1}{4}$ $-\dfrac{7}{12}$
5.	$6\dfrac{3}{5}$ $-2\dfrac{3}{4}$	$4\dfrac{5}{8}$ $-1\dfrac{1}{2}$	$9\dfrac{1}{8}$ $2\dfrac{4}{6}$ $+\dfrac{7}{10}$	$5\dfrac{3}{4}$ $\dfrac{1}{6}$ $+5\dfrac{3}{8}$

Perfect score: 20   My score: _____

PRE-TEST—Division

NAME _____

Chapter 5

Write each answer in simplest form.

	a	b	c	d
1.	$4 \div \frac{1}{2}$	$7 \div \frac{2}{3}$	$8 \div \frac{4}{5}$	$9 \div \frac{6}{7}$
2.	$\frac{1}{4} \div 2$	$\frac{3}{5} \div 2$	$\frac{3}{7} \div 3$	$\frac{8}{9} \div 6$
3.	$\frac{1}{7} \div \frac{1}{2}$	$\frac{1}{8} \div \frac{1}{4}$	$\frac{1}{4} \div \frac{1}{8}$	$\frac{1}{6} \div \frac{1}{7}$
4.	$\frac{1}{8} \div \frac{1}{10}$	$\frac{3}{5} \div \frac{2}{3}$	$\frac{4}{7} \div \frac{2}{7}$	$\frac{5}{6} \div \frac{5}{8}$
5.	$7\frac{1}{2} \div 10$	$3 \div 1\frac{1}{5}$	$\frac{1}{4} \div 1\frac{1}{6}$	$2\frac{1}{2} \div 1\frac{1}{6}$

Perfect score: 20   My score: _____

Lesson 1   Reciprocals

The product of any number and its **reciprocal** is 1.

$$\frac{2}{3} \times \frac{3}{2} = \frac{2 \times 3}{3 \times 2} = \frac{6}{6} = 1$$

The reciprocal of $\frac{2}{3}$ is $\frac{3}{2}$.

The reciprocal of $\frac{3}{2}$ is ____.

$$\frac{1}{2} \times \frac{2}{1} = \frac{1 \times 2}{2 \times 1} = \frac{2}{2} = 1$$

The reciprocal of $\frac{1}{2}$ is $\frac{2}{1}$ or 2.

The reciprocal of 2 is ____.

Write the reciprocal of each of the following.

	a	b	c	d	e	f
1.	$\frac{3}{5}$	$\frac{7}{8}$	$\frac{4}{5}$	$\frac{5}{7}$	$\frac{4}{9}$	$\frac{6}{7}$
2.	$\frac{5}{3}$	$\frac{8}{7}$	$\frac{5}{4}$	$\frac{7}{5}$	$\frac{9}{4}$	$\frac{7}{6}$
3.	$\frac{1}{8}$	$\frac{1}{3}$	$\frac{1}{4}$	$\frac{1}{9}$	$\frac{1}{16}$	$\frac{1}{14}$
4.	$\frac{8}{1}$	$\frac{3}{1}$	$\frac{4}{1}$	$\frac{9}{1}$	$\frac{16}{1}$	$\frac{14}{1}$
5.	8	3	4	9	16	14
6.	$\frac{8}{5}$	6	$\frac{2}{3}$	$\frac{11}{6}$	$\frac{7}{4}$	12
7.	15	$\frac{10}{9}$	$\frac{12}{11}$	17	$\frac{8}{9}$	$\frac{17}{2}$
8.	$\frac{15}{8}$	$\frac{5}{12}$	11	$\frac{7}{11}$	$\frac{1}{11}$	$\frac{17}{3}$
9.	$\frac{10}{1}$	13	$\frac{1}{17}$	$\frac{5}{11}$	$\frac{9}{7}$	5
10.	$\frac{5}{8}$	$\frac{1}{6}$	7	$\frac{12}{7}$	2	$\frac{2}{5}$

Perfect score: 60   My score: _____

Lesson 2   Division

NAME _____

$$15 \div \frac{3}{4} = \frac{15}{1} \times \frac{4}{3}$$
$$= \frac{15 \times 4}{1 \times 3}$$
$$= \frac{60}{3}$$
$$= 20$$

To divide by a fraction, multiply by its reciprocal.

Multiply the fractions.

Write the answer in simplest form.

$$10 \div \frac{6}{7} = \frac{10}{1} \times \frac{7}{6}$$
$$= \frac{10 \times 7}{1 \times 6}$$
$$= \frac{70}{6}$$
$$= 11\frac{2}{3}$$

Write each answer in simplest form.

	a	b	c	d
1.	$10 \div \frac{1}{3}$	$8 \div \frac{1}{2}$	$7 \div \frac{1}{4}$	$6 \div \frac{1}{5}$
2.	$14 \div \frac{2}{7}$	$15 \div \frac{2}{5}$	$16 \div \frac{3}{8}$	$18 \div \frac{5}{9}$
3.	$18 \div \frac{1}{3}$	$14 \div \frac{7}{8}$	$17 \div \frac{1}{2}$	$12 \div \frac{3}{4}$

Perfect score: 12   My score: _____

NAME _____

Lesson 3 Division

$\frac{1}{2} \div 4 = \frac{1}{2} \times \frac{1}{4}$
$= \frac{1 \times 1}{2 \times 4}$
$= \frac{1}{8}$

To divide by a whole number, multiply by its reciprocal.

Multiply the fractions.

$\frac{2}{3} \div 5 = \frac{2}{3} \times \frac{1}{5}$
$= \frac{2 \times 1}{3 \times 5}$
$= \frac{2}{15}$

To divide $\frac{1}{2}$ by 4, multiply $\frac{1}{2}$ by _____.

To divide $\frac{2}{3}$ by 5, multiply $\frac{2}{3}$ by _____.

Write each answer in simplest form.

	a	b	c	d
1.	$\frac{1}{2} \div 6$	$\frac{1}{4} \div 2$	$\frac{1}{3} \div 5$	$\frac{1}{6} \div 2$
2.	$\frac{3}{5} \div 4$	$\frac{5}{8} \div 2$	$\frac{3}{4} \div 4$	$\frac{5}{6} \div 3$
3.	$\frac{3}{4} \div 6$	$\frac{2}{3} \div 6$	$\frac{4}{5} \div 4$	$\frac{5}{6} \div 10$

Perfect score: 12   My score: _____

Problem Solving

Solve. Write each answer in simplest form.

1. One third pound of flour is separated into 2 bowls. The same amount of flour is in each bowl. How much flour is in each bowl?

_____ pound is in each bowl.

2. One half of a room is painted. Each of 4 people did the same amount of painting. How much of the room did each person paint?

Each person painted _____ of the room.

3. Kevin used $\frac{3}{4}$ gallon of gasoline to mow a lawn 3 times. How much gasoline did he use to mow the lawn once?

_____ gallon was used.

4. A string $\frac{2}{3}$ yard long is cut into 4 pieces. Each piece is the same length. How long is each piece?

Each piece is _____ yard long.

5. Seven eighths gallon of liquid is poured into 4 containers. Each container has the same amount in it. How much liquid is in each container?

_____ gallon is in each container.

6. The pupils have $\frac{1}{2}$ hour to complete 3 sections of a quiz. They have the same amount of time to do each section. How much time do they have for each section of the quiz?

They have _____ hour to do each section.

Perfect score: 6   My score: _____

Lesson 4 Division

Multiply by the reciprocal.

$$\frac{1}{4} \div \frac{1}{3} = \frac{1}{4} \times \frac{3}{1}$$
$$= \frac{1 \times 3}{4 \times 1}$$
$$= \frac{3}{4}$$

Multiply by the reciprocal.

$$\frac{3}{4} \div \frac{1}{2} = \frac{3}{4} \times \frac{2}{1}$$
$$= \frac{3 \times 2}{4 \times 1}$$
$$= \frac{6}{4}$$
$$= 1\frac{2}{4}$$
$$= 1\frac{1}{2}$$

Write the answer in simplest form.

Write each answer in simplest form.

	a	b	c	d
1.	$\frac{1}{5} \div \frac{1}{2}$	$\frac{1}{3} \div \frac{1}{2}$	$\frac{1}{8} \div \frac{1}{4}$	$\frac{1}{9} \div \frac{1}{6}$
2.	$\frac{3}{5} \div \frac{1}{2}$	$\frac{4}{7} \div \frac{2}{3}$	$\frac{4}{5} \div \frac{1}{10}$	$\frac{5}{6} \div \frac{2}{3}$
3.	$\frac{4}{5} \div \frac{2}{5}$	$\frac{3}{8} \div \frac{3}{4}$	$\frac{4}{9} \div \frac{1}{5}$	$\frac{7}{8} \div \frac{7}{10}$

Perfect score: 12   My score: _____

Problem Solving

Solve. Write each answer in simplest form.

1. How many $\frac{1}{6}$-hour sessions are there in $\frac{1}{2}$ hour?

   There are _____ sessions.

2. Millie has a ribbon $\frac{3}{4}$ yard long. How many $\frac{1}{4}$-yard pieces can she get from her ribbon?

   She can get _____ pieces.

3. In problem 2, how many $\frac{1}{8}$-yard pieces can Millie get from her ribbon?

   She can get _____ pieces.

4. A machine uses gas at the rate of $\frac{1}{5}$ gallon an hour. So far $\frac{9}{10}$ gallon has been used. How many hours has the machine operated?

   The machine operated _____ hours.

5. Suppose the machine in problem 4 has used $\frac{4}{5}$ gallon of gas. How many hours did the machine operate?

   The machine operated _____ hours.

6. Three eighths pound of nuts is put in each bag. How many bags can be filled with $\frac{3}{4}$ pound of nuts?

   _____ bags can be filled.

7. Wilbur walked $\frac{5}{6}$ hour. He walked at the rate of 1 mile every $\frac{1}{6}$ hour. How many miles did he walk?

   He walked _____ miles.

8. Suppose in problem 7 Wilbur walked 1 mile every $\frac{5}{12}$ hour. How many miles did he walk?

   He walked _____ miles.

9. A bell rings every $\frac{1}{6}$ hour. Assume it just rang. How many times will it ring in the next $\frac{2}{3}$ hour?

   It will ring _____ times.

Perfect score: 9   My score: _____

Lesson 5 Division

Write each answer in simplest form.

	a	b	c	d
1.	$8 \div \frac{2}{3}$	$\frac{4}{7} \div 5$	$\frac{1}{6} \div \frac{1}{3}$	$\frac{3}{5} \div \frac{2}{3}$
2.	$\frac{1}{8} \div \frac{1}{10}$	$6 \div \frac{1}{4}$	$\frac{1}{3} \div 2$	$\frac{1}{7} \div \frac{1}{3}$
3.	$\frac{1}{2} \div \frac{1}{5}$	$\frac{9}{10} \div \frac{4}{5}$	$9 \div \frac{3}{5}$	$\frac{4}{9} \div 6$
4.	$\frac{3}{5} \div 3$	$\frac{1}{3} \div \frac{1}{6}$	$\frac{3}{8} \div \frac{3}{10}$	$6 \div \frac{4}{5}$
5.	$\frac{7}{8} \div \frac{7}{8}$	$\frac{6}{7} \div \frac{8}{9}$	$7 \div \frac{1}{3}$	$\frac{6}{7} \div 4$

Perfect score: 20   My score: _____

Problem Solving

Solve. Write each answer in simplest form.

1. It takes $\frac{1}{3}$ hour to produce 1 woomble. How many woombles could be produced in 9 hours?

_____ woombles could be produced.

2. A rope $\frac{3}{4}$ yard long is cut into 9 pieces. Each piece is the same length. How long is each piece?

Each piece is _____ yard long.

3. Each class period is $\frac{3}{5}$ hour long. How many topics can be covered in 1 class period if it takes $\frac{3}{10}$ hour to cover each topic?

_____ topics can be covered.

4. Eight pounds of raisins are put in boxes. How many boxes are needed if $\frac{2}{3}$ pound of raisins is put into each box?

_____ boxes are needed.

5. Michelle and her 5 friends want to share $\frac{3}{4}$ pound of salami equally. How much salami will each person get?

Each person will get _____ pound.

6. A wire $4\frac{1}{2}$ feet long is cut into 9 pieces of the same length. How long is each piece?

Each piece is _____ foot long.

7. Three fourths gallon of milk is poured into 12 glasses. The same amount is in each glass. How much milk is in each glass?

_____ gallon is in each glass.

8. Mr. Roe has $\frac{9}{10}$ pound of a chemical to put into 6 tubes. Assume he puts the same amount in each tube. How many pounds of chemical will be in each tube?

_____ pound will be in each tube.

Perfect score: 8   My score: _____

Lesson 6 Division

$2\frac{1}{5} \div 4 = \frac{11}{5} \div 4$ Change the mixed numerals to fractions.

$= \frac{11}{5} \times \frac{1}{4}$ To divide, multiply by the reciprocal.

$= \frac{11}{20}$ Multiply the fractions.

Write the answer in simplest form.

$3\frac{1}{2} \div 1\frac{1}{2} = \frac{7}{2} \div \frac{3}{2}$

$= \frac{7}{2} \times \frac{2}{3}$

$= \frac{14}{6}$

$= 2\frac{1}{3}$

Write each answer in simplest form.

	a	b	c	d
1.	$2\frac{1}{2} \div 3$	$1\frac{2}{5} \div 3$	$4 \div 1\frac{1}{3}$	$6 \div 1\frac{1}{3}$
2.	$1\frac{2}{7} \div 2\frac{1}{2}$	$1\frac{1}{5} \div 2\frac{2}{3}$	$4\frac{1}{2} \div 1\frac{1}{5}$	$1\frac{4}{5} \div 1\frac{1}{5}$
3.	$1\frac{4}{5} \div \frac{2}{7}$	$\frac{1}{6} \div 1\frac{1}{2}$	$3\frac{3}{5} \div 10$	$1\frac{1}{3} \div 2\frac{1}{2}$

Perfect score: 12   My score: _____

Problem Solving

Solve. Write each answer in simplest form.

1. Five pounds of sand are put into containers. How many containers are needed if $1\frac{1}{4}$ pounds of sand are put into each one?

   _____ containers are needed.

2. Yvonne works $1\frac{1}{2}$ hours each day. How many days will it take her to work 15 hours?

   It will take _____ days.

3. Each class period is $\frac{5}{6}$ hour long. How many periods can there be in $2\frac{1}{2}$ hours?

   There can be _____ periods in $2\frac{1}{2}$ hours.

4. The city spread $7\frac{1}{2}$ tons of salt on the streets. There were $1\frac{1}{4}$ tons on each load. How many loads of salt were spread on the streets?

   _____ loads of salt were spread.

5. It takes $1\frac{5}{6}$ hours to assemble a lawn mower. How many lawn mowers could be assembled in $16\frac{1}{2}$ hours?

   _____ lawn mowers could be assembled.

6. How many $1\frac{1}{2}$-hour practice sessions are there in 6 hours?

   There are _____ practice sessions.

7. How many $1\frac{3}{4}$-hour practice sessions are there in $10\frac{1}{2}$ hours?

   There are _____ practice sessions.

1.
2.
3.
4.
5.
6.
7.

Perfect score: 7   My score: _____

CHAPTER 5 TEST

Write each answer in simplest form.

	a	b	c	d
1.	$5 \div \dfrac{1}{3}$	$8 \div \dfrac{3}{4}$	$4 \div \dfrac{2}{3}$	$10 \div \dfrac{6}{7}$
2.	$\dfrac{1}{2} \div 3$	$\dfrac{4}{7} \div 3$	$\dfrac{5}{9} \div 5$	$\dfrac{6}{7} \div 8$
3.	$\dfrac{1}{9} \div \dfrac{1}{4}$	$\dfrac{1}{10} \div \dfrac{1}{5}$	$\dfrac{1}{5} \div \dfrac{1}{10}$	$\dfrac{1}{3} \div \dfrac{1}{4}$
4.	$\dfrac{1}{6} \div \dfrac{1}{9}$	$\dfrac{4}{5} \div \dfrac{3}{4}$	$\dfrac{7}{9} \div \dfrac{2}{3}$	$\dfrac{5}{8} \div \dfrac{5}{6}$
5.	$2\dfrac{1}{3} \div 5$	$6 \div 1\dfrac{2}{3}$	$\dfrac{1}{3} \div 1\dfrac{1}{2}$	$3\dfrac{1}{3} \div 1\dfrac{1}{2}$

Perfect score: 20   My score: _____

PRE-TEST—Addition and Subtraction

NAME _____

Chapter 6

Change each fraction or mixed numeral to a decimal.

	a	b	c
1.	$\frac{7}{10} =$ _____	$3\frac{19}{100} =$ _____	$5\frac{25}{1000} =$ _____

Change each of the following to a decimal as indicated.

2. Change $\frac{4}{5}$ to tenths.   Change $3\frac{8}{25}$ to hundredths.   Change $3\frac{16}{125}$ to thousandths.

Change each decimal to a fraction or mixed numeral in simplest form.

	a	b	c
3.	.8	9.33	16.125

Add or subtract.

	a	b	c	d
4.	.6 +.2	.7 2 +.3 5	5.3 8 2 +2.6 4 1	5.0 1 8 3.2 4 6 +5.8 1 2
5.	4.6 −3.5	5.0 5 −4.2 9	.4 5 6 −.0 1 8	1 2.0 3 8 −1.7 6 4
6.	.7 +.3 8	.2 5 6 +.8 3	1.7 +5.8 2 5	3.4 4 2.0 1 8 +.7 9
7.	.4 2 −.1	4.5 6 −1.2 4 3	5.8 −2.2 5	1 6.3 6 −1 6.0 7 5

Perfect score: 25   My score: _____

Lesson 1 Tenths

NAME _____

Numerals like .4, 4.1, and 5.4 are called **decimals**.

$\frac{1}{10}$ = .1   .1 is read "one tenth."

decimal points

.4 = $\frac{4}{10}$ $\frac{3}{10}$ = .3

$4\frac{1}{10}$ = 4.1   4.1 is read "four and one tenth."

5.4 = ____ $2\frac{3}{10}$ = ____

Change each fraction or mixed numeral to a decimal.

	a	b	c	d
1.	$\frac{6}{10}$ = ____	$\frac{2}{10}$ = ____	$\frac{8}{10}$ = ____	$\frac{5}{10}$ = ____
2.	$4\frac{7}{10}$ = ____	$5\frac{9}{10}$ = ____	$18\frac{2}{10}$ = ____	$423\frac{6}{10}$ = ____

Change each decimal to a fraction or mixed numeral.

3. .7 = ____   .3 = ____   .1 = ____   .9 = ____

4. 4.9 = ____   12.7 = ____   15.1 = ____   217.3 = ____

Write a decimal for each of the following.

        a   b

5. eight tenths _____   three and seven tenths _____

6. four tenths _____   twenty-five and eight tenths _____

7. five tenths _____   one hundred and six tenths _____

Write each decimal in words.

8. .9 _____

9. 3.7 _____

10. 21.2 _____

Perfect score: 25   My score: _____

Lesson 2 Hundredths

NAME _____

$\frac{1}{100} = .01$ .01 is read "one hundredth."

$.15 = \frac{15}{100}$ $\frac{9}{100} = .09$

$3\frac{12}{100} = 3.12$ 3.12 is read "three and twelve hundredths." $2.07 = $ _____ $1\frac{14}{100} = $ _____

Change each fraction or mixed numeral to a decimal naming hundredths.

　　　　　　　　　　　a　　　　　　　　　　　　　b　　　　　　　　　　　　　c

1.　　$\frac{8}{100} = $ _____　　　　$\frac{16}{100} = $ _____　　　　$\frac{5}{100} = $ _____

2.　　$1\frac{36}{100} = $ _____　　　　$8\frac{6}{100} = $ _____　　　　$9\frac{12}{100} = $ _____

3.　　$12\frac{45}{100} = $ _____　　　　$43\frac{67}{100} = $ _____　　　　$26\frac{4}{100} = $ _____

4.　　$142\frac{8}{100} = $ _____　　　　$436\frac{42}{100} = $ _____　　　　$389\frac{89}{100} = $ _____

Change each decimal to a fraction or mixed numeral.

5.　　.17 = _____　　　　.03 = _____　　　　.41 = _____

6.　　5.19 = _____　　　　6.47 = _____　　　　5.01 = _____

7.　　21.07 = _____　　　　23.99 = _____　　　　44.89 = _____

8.　　142.33 = _____　　　　483.03 = _____　　　　185.63 = _____

Write a decimal for each of the following.

　　　　　　　　　a　　　　　　　　　　　　　　　　　　b

9.　eight hundredths　_____　　　six and twenty-three hundredths　_____

10.　ninety-five hundredths　_____　　　fourteen and sixty hundredths　_____

11.　forty-eight hundredths　_____　　　four and forty-four hundredths　_____

Perfect score: 30   My score: _____

NAME _____

Lesson 3   Thousandths, Ten Thousandths

$\frac{1}{1000} = .001$ ← one thousandth

$2\frac{12}{1000} = 2.012$ ← two and twelve thousandths

$\frac{1}{10000} = .0001$ ← one ten thousandth

$1\frac{35}{10000} = 1.0035$ ← one and thirty-five ten thousandths

---

Write each fraction or mixed numeral as a decimal.

	a	b	c
1.	$\frac{8}{1000} =$	$\frac{17}{1000} =$	$\frac{54}{10000} =$
2.	$\frac{125}{10000} =$	$\frac{430}{1000} =$	$\frac{306}{10000} =$
3.	$4\frac{4}{1000} =$	$3\frac{41}{10000} =$	$6\frac{183}{1000} =$
4.	$35\frac{78}{10000} =$	$42\frac{19}{1000} =$	$196\frac{6}{1000} =$

Write each decimal as a fraction or as a mixed numeral.

	a	b	c
5.	.009 =	.0019 =	.0003 =
6.	.123 =	.0441 =	.219 =
7.	4.011 =	2.1011 =	6.0014 =
8.	36.037 =	3.433 =	100.0001 =

Write a decimal for each of the following.

	a	b
9.	fifty-three thousandths _____	ten and nine ten thousandths _____
10.	eleven ten thousandths _____	twelve and eighteen thousandths _____
11.	sixty-five thousandths _____	twelve and one thousandth _____

Perfect score: 30   My score: _____

Lesson 4 Changing Fractions and Mixed Numerals to Decimals

Change $\frac{1}{2}$ to tenths.

$$\frac{1}{2} = \frac{1}{2} \times \frac{5}{5}$$
$$= \frac{5}{10}$$
$$= .5$$

Change $\frac{1}{2}$ to hundredths.

$$\frac{1}{2} = \frac{1}{2} \times \frac{50}{50}$$
$$= \frac{50}{100}$$
$$= .50$$

Change $\frac{1}{2}$ to thousandths.

$$\frac{1}{2} = \frac{1}{2} \times \frac{500}{500}$$
$$= \frac{500}{1000}$$
$$= .500$$

Change $\frac{3}{4}$ to hundredths.

$$\frac{3}{4} = \frac{3}{4} \times \frac{25}{25}$$
$$= \frac{75}{100}$$
$$= \underline{\hspace{2cm}}$$

Change $3\frac{48}{250}$ to thousandths.

$$3\frac{48}{250} = 3 + \frac{48}{250}$$
$$= 3 + \left(\frac{48}{250} \times \frac{4}{4}\right)$$
$$= 3 + \frac{192}{1000}$$
$$= 3\frac{192}{1000}$$
$$= \underline{\hspace{2cm}}$$

Change each of the following to a decimal as indicated.

	a	b	c
1.	Change $\frac{3}{5}$ to tenths.	Change $\frac{3}{5}$ to hundredths.	Change $\frac{3}{5}$ to thousandths.
2.	Change $3\frac{1}{2}$ to tenths.	Change $\frac{7}{25}$ to hundredths.	Change $2\frac{19}{100}$ to thousandths.
3.	Change $2\frac{4}{5}$ to tenths.	Change $\frac{7}{20}$ to hundredths.	Change $\frac{7}{125}$ to thousandths.
4.	Change $2\frac{1}{5}$ to tenths.	Change $\frac{19}{50}$ to hundredths.	Change $\frac{88}{250}$ to thousandths.

Perfect score: 12   My score: _____

NAME _____

Lesson 5 Changing Decimals to Fractions or Mixed Numerals

$.7 = \frac{7}{10}$
$.19 = \frac{19}{100}$

$.6 = \frac{6}{10}$ or $\frac{3}{5}$
$.14 = \frac{14}{100}$ or $\frac{7}{50}$

$4.2 = 4\frac{2}{10}$ or $4\frac{1}{5}$
$3.01 = 3\frac{1}{100}$

$.051 = $ _____

$.114 = \frac{114}{1000}$ or _____

$5.006 = 5\frac{6}{1000}$ or _____

Change each decimal to a fraction or mixed numeral in simplest form.

	a	b	c	d
1.	.3	.1	.4	.5
2.	2.7	3.3	7.2	5.8
3.	.17	.03	.15	.80
4.	5.07	8.43	4.05	2.44
5.	.003	.017	.125	.045
6.	3.121	2.987	4.250	3.008
7.	4.35	.7	6.200	1.007
8.	2.6	3.24	.250	3.5
9.	5.125	.9	2.4	.04
10.	.01	.051	.8	2.19

Perfect score: 40   My score: _____

Lesson 6 Fractions, Mixed Numerals, and Decimals

Change each of the following to a decimal as indicated.

	a	b	c
1.	Change $\frac{1}{5}$ to tenths.	Change $\frac{7}{20}$ to hundredths.	Change $\frac{89}{200}$ to thousandths.
2.	Change $7\frac{1}{2}$ to tenths.	Change $4\frac{29}{50}$ to hundredths.	Change $3\frac{9}{25}$ to thousandths.

Change each decimal to a fraction or mixed numeral in simplest form.

	a	b	c	d
3.	.9	3.6	.35	17.75
4.	.025	8.445	24.305	8.05

Complete the following so the numerals in each row name the same number.

| | fractions or mixed numerals | decimals | | |
		tenths	hundredths	thousandths
5.				.600
6.			2.70	
7.		5.4		
8.	$3\frac{1}{2}$			
9.		17.9		
10.			80.80	

Perfect score: 32   My score: _____

Lesson 7 Addition

When adding decimals, line up the decimal points. Add decimals like you add whole numbers.

```
          3.⁵6              3.0¹8
   .6     .03               .142
 +.7    +4.24             +14.009
 ─────   ──────            ────────
  1.3     7.83             17.169
```
← Place the decimal point in the answer. →

Add.

	a	b	c	d	e
1.	.4 +.5	.9 +.8	3.4 +9.2	19.3 +12.8	45.6 +6.8
2.	.42 +.35	.76 +.48	3.32 +4.62	24.45 +72.36	58.92 +3.29
3.	.014 +.231	.456 +.876	2.014 +2.325	3.457 +2.356	41.216 +2.007
4.	.5 .6 +.7	1.9 2.2 +3.4	3.4 1.7 +4.8	42.3 1.6 +2.9	3.4 .8 +4.2
5.	.33 .26 +.41	$.43 .54 +.07	3.35 1.08 +6.11	$24.29 12.29 +5.31	$34.05 2.06 +1.08
6.	.012 .304 +.405	.423 .056 +.217	3.056 1.452 +6.112	4.008 2.309 +.012	35.157 .448 +2.509

Perfect score: 30   My score: _____

Problem Solving

Solve each problem.

1. There was .8 inch of rain recorded on Monday, .5 inch on Tuesday, and .7 inch on Friday. How many inches of rain were recorded on those 3 days?

    _____ inches were recorded.

2. In problem 1, how many inches of rain were recorded on Monday and Friday?

    _____ inches were recorded.

3. A board is 4.25 meters long. Another is 3.75 meters long. When laid end to end, the boards are how long?

    The boards are _____ meters long.

4. One sheet of metal is .28 centimeter thick. Another is .35 centimeter thick. What would be the combined thickness of these sheets?

    The thickness would be _____ centimeter.

5. Three sheets of metal are to be placed on top of each other. Their thicknesses are .125 inch, .018 inch, and .075 inch. What would be the combined thickness of all three pieces?

    The combined thickness would be _____ inch.

6. Box A weighs 1.4 pounds, box B weighs 3.2 pounds, and box C weighs 2.5 pounds. What is the combined weight of box A and box C?

    The combined weight is _____ pounds.

7. In problem 6, what is the combined weight of all three boxes?

    The combined weight is _____ pounds.

8. Mary Ann made three purchases at the store. The amounts were $13.75; $1.42; and $.83. What was the total amount of all three purchases?

    The total amount was $_____.

Perfect score: 8   My score: _____

Lesson 8 Addition

NAME _____

You may write these 0's if they help you add.

```
   .8          .8 0              4.2          4.2 0 0
 +.3 9   or  +.3 9             3.0 1 8      3.0 1 8
 -----       -----             +.8 2    or  +.8 2 0
  1.1 9       1.1 9            -------      -------
                                8.0 3 8      8.0 3 8
```

Add. If necessary, use 0's as shown in the examples.

	a	b	c	d	e
1.	.9 +.4 2	.8 3 +.4	.6 +.4 0 1	.7 2 +.4 2 3	.6 4 5 +.2
2.	2.7 5 +3.3 0 8	5.5 4 +7.6	3.8 +.3 1 6	.2 9 +8.0 4 3	2 9.5 +4.9 3
3.	.4 2 .8 +.0 1 8	.3 1 .2 +.4 5	.7 6 .8 2 +.9	.4 3 1 .2 +.4 5	.5 .3 1 6 +.0 9 9
4.	3.1 8 2 1.3 4 +2.6	4.7 2 5.8 +6.3 1 7	7.4 2 6 3.3 1 8 +.2	.7 3 1 8.4 5 +2.2 8	.3 .3 8 4 +9.4 2

Complete the following.

	a	b
5.	.8 + .91 = _____	.4 + .016 + .75 = _____
6.	.58 + .114 = _____	.32 + .42 + .113 = _____
7.	.9 + .301 = _____	4.8 + 3.21 + .014 = _____
8.	2.4 + .31 = _____	5.24 + .016 + 21.3 = _____

Perfect score: 28   My score: _____

Problem Solving

Solve each problem.

1. There was .75 inch of rain recorded at Elmhurst, .50 inch at River Forest, and .25 inch at Harvey. What amount of rain was recorded at both Elmhurst and River Forest?

   The total amount was _____ inches of rain.

2. In problem 1, how much rain was recorded at all three locations?

   _____ inches were recorded.

3. An opening in an engine part is supposed to be 1.150 centimeters. The part is acceptable if the opening is as much as .075 centimeter larger than what it is supposed to be. What is the largest opening that would be acceptable?

   The largest opening is _____ centimeters.

4. Assume the opening in problem 3 can only be as much as .025 centimeter larger than what it is supposed to be. What is the largest acceptable opening?

   The largest opening is _____ centimeters.

5. Andy saved $23.05. Betty saved $40. Clare saved $3.50. How much have all three saved?

   All three have saved a total of _____.

6. In problem 5, how much have Andy and Clare saved?

   They have saved _____.

7. In problem 5, how much have Betty and Clare saved?

   They have saved _____.

8. Marlene was asked to find the sum of 1.9; 3.52; and .075. What should her answer be?

   Her answer should be _____.

Perfect score: 8   My score: _____

Lesson 9   Subtraction

NAME _____

When subtracting decimals, line up the decimal points. Subtract decimals like you subtract whole numbers.

```
         3 13              0 14         3 12   4 13
   9.5     4.3              .1 4         4 2.7 5 3
 − 2.3   − 1.6            − .0 8       − 5.3 2 7
 ─────   ─────            ──────       ─────────
   7.2     2.7              .0 6         3 7.4 2 6
```

Place the decimal point in the answer.

Subtract.

	a	b	c	d	e
1.	.7 −.3	.9 −.2	.6 −.2	.9 −.1	.8 −.5
2.	.4 2 −.3 1	.5 6 −.2 3	.0 7 −.0 2	.8 5 −.3 7	$ .5 2 −.3 7
3.	.3 4 5 −.2 3 4	.5 4 8 −.2 5 9	.8 1 5 −.6 0 7	.8 2 8 −.3 8 9	.7 5 4 −.3 7 5
4.	4.6 −3.2	7.4 −2.8	8.6 −3.7	5.6 −.7	1 9.2 −.9
5.	4.3 6 −1.2 3	$ 6.5 5 −2.7 3	4.0 8 −.3 9	$ 1 5.3 2 −2.6 7	$ 4.0 9 −.3 2
6.	4.2 1 3 −2.0 0 1	3.6 2 4 −1.4 1 5	4.3 0 7 −1.4 9 5	2 6.3 4 5 −2.5 4 3	1 5.1 0 8 −3.9 1 2
7.	1 5.3 −4.9	6.2 3 −3.7 5	1 4.2 1 −7.0 8	3.0 0 2 −1.0 4 7	1 9.8 0 1 −7.4 1 3

Perfect score: 35   My score: _____

Problem Solving

Solve each problem.

1. Fran is to mix .8 pound of chemical A, .6 pound of chemical B, and .3 pound of chemical C. How much more of chemical A is to be used than chemical B?

   _____ pound more of chemical A is to be used.

2. In problem 1, how much more of chemical A than chemical C is to be used?

   _____ pound more of chemical A is to be used.

3. A spark plug has a gap of 1.12 millimeters. The gap should be .89 millimeter. How much too large is the gap?

   It is _____ millimeter too large.

4. Suppose the gap in problem 3 was .65 millimeter. How much too small is the gap?

   It is _____ millimeter too small.

5. Three sheets of metal were placed together. Their total thickness was 4.525 inches. Then a sheet 1.750 inches thick was removed. What was the combined thickness of the remaining sheets?

   It is _____ inches thick.

6. The distance between two terminals on a television part is supposed to be 2.45 inches. The part is acceptable if the distance is .05 inches more or less than what it is supposed to be. What is the least distance that would be acceptable?

   The least distance would be _____ inches.

7. One box of nails weighs 3.4 pounds and another box weighs 5.2 pounds. How much more does the heavier box weigh?

   The heavier box weighs _____ pounds more.

Perfect score: 7   My score: _____

Lesson 10   Subtraction

NAME _____

$\begin{array}{r}\overset{5\ 14}{6.4}3\ 2\\-1.7\\\hline 4.7\ 3\ 2\end{array}$ or $\begin{array}{r}\overset{5\ 14}{6.4}3\ 2\\-1.7\ 0\ 0\\\hline 4.7\ 3\ 2\end{array}$ ← Write these 0's if they help you.

$\begin{array}{r}6.4\\-1.2\ 3\\\hline\end{array}$ → $\begin{array}{r}\overset{3\ 10}{6.4}\ \cancel{0}\\-1.2\ 3\\\hline 5.1\ 7\end{array}$ ← Write this 0 to help you subtract.

Subtract.

	a	b	c	d	e
1.	.72 −.2	3.56 −1.4	5.38 −2.7	4.316 −1.1	2.146 −1.5
2.	.523 −.41	.683 −.39	5.421 −.56	3.018 −.27	4.012 −3.03
3.	.8 −.35	.5 −.26	6.3 −1.12	7.4 −2.75	14.3 −6.72
4.	.9 −.309	.3 −.175	4.4 −2.356	6.3 −3.432	18.2 −7.514
5.	.75 −.314	.36 −.275	5.72 −1.312	4.38 −.592	16.92 −6.384
6.	34.265 −2.18	42.16 −3.235	42.2 −3.164	26.3 −2.45	3.106 −2.03
7.	43.7 −6.18	394.6 −75.81	5.216 −4.19	82.45 −3.783	92.405 −3.008

Perfect score: 35   My score: _____

Problem Solving

Today's Work Report		
Ms. Williams	14.7 units	1.2 hours
Mr. Karns	8.4 units	.9 hour
Mr. Anders	13.5 units	1.4 hours

The manufacturing director uses her computer to find out how many units her workers are producing. Use the information above to solve each problem.

1. How many more units did Ms. Williams make than Mr. Anders?

   Ms. Williams made _____ more units.

2. Who made the most units? Who made the fewest units? What is the difference between the most and the fewest units made?

   _____ made the most units.

   _____ made the fewest units.

   The difference is _____ units.

3. How many units did the three workers make in all?

   The three workers made _____ units.

4. How long did the three workers work on the units in all?

   The three workers worked _____ hours.

Perfect score: 6   My score: _____

CHAPTER 6 TEST

NAME _____

Change each fraction or mixed numeral to a decimal.

a	b	c
1. $\frac{175}{1000} =$ _____	$9\frac{4}{10} =$ _____	$3\frac{8}{100} =$ _____

Change each of the following to a decimal as indicated.

2. Change $\frac{9}{10}$ to hundredths.   Change $3\frac{1}{5}$ to tenths.   Change $5\frac{75}{250}$ to thousandths.

Change each decimal to a fraction or mixed numeral in simplest form.

a	b	c
3. .075	8.6	16.49

Add or subtract.

	a	b	c	d
4.	.9 +.4	.52 +.43	6.534 +7.827	9.308 21.295 +.043
5.	3.3 −1.6	8.24 −3.73	.442 −.375	18.042 −12.345
6.	.42 +.9	.35 +.065	3.6 +14.673	9.2 4.375 +43.78
7.	.546 −.38	3.8 −1.21	7.22 −4.436	8.4 −3.575

Perfect score: 25   My score: _____

PRE-TEST—Multiplication

NAME _____

Chapter 7

Multiply.

	a	b	c	d	e
1.	.7 ×5	.4 ×2	4 ×.9	4 ×.3	5 ×.6
2.	.07 ×6	.02 ×3	8 ×.04	8 ×.09	5 ×.08
3.	.003 ×4	.001 ×8	2 ×.003	4 ×.001	6 ×.002
4.	.7 ×.3	.2 ×.4	.4 ×.6	.9 ×.6	.8 ×.5
5.	.06 ×.8	.01 ×.5	.2 ×.03	.06 ×.07	.02 ×.02
6.	.4 ×10	.008 ×100	6.72 ×10	.234 ×1000	5.68 ×1000
7.	16 ×.3	.47 ×.5	3.4 ×.08	5.01 ×.25	.078 ×7.5

Perfect score: 35   My score: _____

Lesson 1 Multiplication

NAME _____

number of digits to the right of the
decimal point

```
    4        .4    1      .04    2       .04    2       .04    2
   ×3       ×3    +0      ×3    +0      ×.3   +1      ×.03   +2
   ──       ──    ──      ──    ──      ───    ─      ────    ─
   12       1.2    1      .12    2      .012    3     .0012    4
```

Write in as many 0's as needed to place the decimal point correctly.

Multiply.

	a	b	c	d	e
1.	2 ×3	.2 ×3	.02 ×3	.002 ×3	2 ×.3
2.	8 ×6	.8 ×6	.08 ×6	.008 ×6	.06 ×8
3.	5 ×3	.5 ×3	.05 ×3	.005 ×3	.003 ×5
4.	3 ×4	.3 ×.4	.03 ×.4	.04 ×.3	.03 ×.04
5.	6 ×7	.6 ×.7	.06 ×.7	.07 ×.6	.06 ×.07
6.	9 ×8	.9 ×.8	.09 ×.8	.08 ×.9	.09 ×.08

Perfect score: 30   My score: _____

83

Lesson 2 Multiplication

NAME _____

number of digits to the right of the
decimal point

```
  24        2.4     1      .24    2      .24    2      .24    2
× 36      × 36    +0     × 36   +0     ×3.6   +1     ×.36   +2
────      ────    ──     ────   ──     ────   ──     ────   ──
 864      86.4     1      8.64   2     .864    3    .0864    4
```

Use the completed multiplication to find each product.

		a	b	c	d
1.	32 ×14 ──── 448	3.2 ×14	.32 ×14	.32 ×1.4	.32 ×.14
2.	27 ×48 ──── 1296	2.7 ×48	.27 ×48	.27 ×4.8	.27 ×.48
3.	26 ×34 ──── 884	.26 ×34	.26 ×3.4	.26 ×.34	2.6 ×34
4.	74 ×26 ──── 1924	.74 ×2.6	7.4 ×26	.74 ×26	.74 ×.26
5.	25 ×3 ─── 75	25 ×.3	2.5 ×.03	25 ×.03	.25 ×.03
6.	12 ×4 ─── 48	1.2 ×.4	.12 ×4	.12 ×.4	.12 ×.04
7.	73 ×3 ─── 219	73 ×.03	.73 ×.03	7.3 ×.3	.73 ×.3

Perfect score: 28   My score: _____

Lesson 3 Multiplication

NAME _____

```
   6        .6        .6       .06       .06       .06      .006
  ×3        ×3       ×.3       ×3       ×.3       ×.03      ×.3
  ──       ──       ──        ──       ────      ─────     ─────
  18       1.8       .18       .18      .018      .0018     .0018
```

Multiply.

	a	b	c	d	e
1.	.7 ×5	3 ×.2	.8 ×9	7 ×.3	.2 ×4
2.	.8 ×.6	.1 ×.6	.3 ×.7	.2 ×.4	.7 ×.6
3.	.08 ×4	.02 ×3	7 ×.08	.09 ×6	5 ×.03
4.	.05 ×.9	.7 ×.05	.08 ×.8	.2 ×.03	.03 ×.5
5.	.03 ×.08	.04 ×.06	.09 ×.01	.07 ×.08	.03 ×.02
6.	.007 ×9	6 ×.008	.004 ×6	.008 ×4	5 ×.007
7.	.005 ×.9	.009 ×.9	.3 ×.004	.003 ×.3	.5 ×.005

Perfect score: 35   My score: _____

Lesson 4 Multiplication

NAME _____

Shortcut

```
  2.51        2.51         2.51
× 10        × 100       × 1000
─────       ──────      ───────
25.10       251.00      2510.00
  or          or           or
 25.1         251         2510
```

$2.51 \times 10 = 25.1$

$2.51 \times 100 = 251$

$2.51 \times 1000 = 2510$

```
  .085        .085         .085
×  10       × 100       × 1000
─────       ──────      ───────
 .850       8.500        85.000
  or          or           or
  .85         8.5          85
```

$.085 \times 10 = 0.85$

$.085 \times 100 = 08.5$

$.085 \times 1000 = 085$

Multiply.

	a	b	c	d	e
1.	5.6 4 2 × 1 0	5.6 4 2 × 1 0 0	5.6 4 2 × 1 0 0 0	5 6.4 2 × 1 0 0	.5 6 4 2 × 1 0
2.	.1 0 6 4 × 1 0	.1 0 6 4 × 1 0 0	.1 0 6 4 × 1 0 0 0	.0 1 0 6 × 1 0	1.0 6 4 × 1 0 0 0
3.	.2 3 × 1 0	.2 3 × 1 0 0	.2 3 × 1 0 0 0	.0 2 3 × 1 0	.0 0 2 3 × 1 0 0
4.	.0 0 8 × 1 0	.0 0 8 × 1 0 0	.0 0 8 × 1 0 0 0	.0 8 × 1 0 0	.0 8 × 1 0 0 0
5.	1.5 × 1 0	1.5 × 1 0 0	1.5 × 1 0 0 0	1 5 × 1 0 0	.1 5 × 1 0

Perfect score: 25   My score: _____

Lesson 5 Multiplication

Multiply as whole numbers. → Place the decimal point in the product.

```
  .356           .356            .356
 ×4.2           ×4.2            ×4.2
                 712             712
                1424            1424
               14952           1.4952
```

Multiply.

1.
   43   5.4   .076   .18   .092
×.08   ×.04   ×.7   ×.09   ×8

2.
.137   4.82   907   6.53   .416
×.3   ×8   ×.4   ×.7   ×6

3.
32.1   5.06   .0709   .421   .0503
×.5   ×4   ×4   ×.2   ×9

4.
.27   5.8   .03   .42   7.6
×4.2   ×.16   ×2.5   ×.53   ×7.6

5.
.187   .084   16.1   .072   5.21
×3.5   ×42   ×5.3   ×6.2   ×.75

6.
42.16   .4218   306.4   .0314   .0144
×1.8   ×22   ×.24   ×26   ×37

Perfect score: 30   My score: _____

Problem Solving

Solve each problem.

1. A box of seeds weighs .9 pound. How many pounds would 6 boxes weigh?

They would weigh _____ pounds.

2. A machinist has 4 sheets of metal, each .042 inch thick. These are placed one on top of the other. What is the total thickness of the sheets?

It will be _____ inch thick.

3. Each needle weighs .03 gram. How many grams would 100 needles weigh?

They would weigh _____ grams.

4. A log weighs 42.1 kilograms. How much would .5 of the log weigh?

It would weigh _____ kilograms.

5. The thickness of a sheet of plastic is .024 inch. What would be the combined thickness of 6 sheets of plastic?

The combined thickness would be _____ inch.

6. In problem 5, what would be the combined thickness of 8 sheets of plastic?

The combined thickness would be _____ inch.

7. Mrs. Tomasello has 92 sheets of foil. Each sheet is .0413 inch thick. What is the combined thickness of the sheets?

The thickness would be _____ inches.

8. Mr. McClean's car averages 5.7 kilometers per liter of gasoline. How many kilometers would he be able to travel with 84 liters of gasoline?

He could travel _____ kilometers.

9. Suppose the car in problem 8 averages 4.9 kilometers per liter. How far could the car go on 84 liters of gasoline?

It could go _____ kilometers.

Perfect score: 9   My score: _____

Lesson 6 Multiplication

Multiply.

	a	b	c	d
1.	.6 ×.4	.08 × 9	.07 ×.07	.005 × 8
2.	4.5 ×.27	.38 ×.32	2.6 ×.043	7.5 ×2.5
3.	.149 ×53	47.6 ×.042	3.08 ×5.3	.729 ×6.1
4.	35.46 ×.27	318.2 ×.36	9.804 ×26	800.6 ×.043
5.	7.21 ×5.34	40.7 ×4.31	312 ×.0624	.598 ×75.3

Perfect score: 20   My score: _____

Problem Solving

Solve each problem.

1. Each box of bolts weighs 1.7 pounds. There are 24 boxes in a carton. How many pounds would a carton of bolts weigh?

A carton would weigh _____ pounds.

2. A sheet of paper is .012 centimeter thick. How many centimeters thick would a stack of paper be if it contained 28 sheets of paper?

The stack would be _____ centimeter thick.

3. Each sheet of metal is .024 centimeter thick. There are 67 sheets of metal in a stack. How high is the stack?

The stack is _____ centimeters high.

4. Mrs. Washington's car averaged 18.36 miles per gallon of gasoline. She bought 11.25 gallons of gasoline. How many miles can she travel on that amount of gasoline?

She can travel _____ miles.

5. After a tune-up, Mrs. Washington's car averaged 21.78 miles per gallon of gasoline. How many miles can she travel on 11.25 gallons of gasoline?

She can travel _____ miles.

6. Each container filled with chemical X weighs 32.7 pounds. How many pounds would 100 containers weigh?

They would weigh _____ pounds.

7. To make each unit takes .035 hour. How long will it take to make 224 units?

It will take _____ hours.

8. Suppose a new machine can make each unit in .018 hour. How long will that machine take to make 224 units?

It will take _____ hours.

Perfect score: 8   My score: _____

CHAPTER 7 TEST

NAME _____

Multiply.

	a	b	c	d	e
1.	.6 ×.8	.18 ×7	.308 ×.9	.42 ×5.3	1.73 ×2.8
2.	9 ×.6	2.4 ×.3	42.6 ×.7	.64 ×.75	146 ×.52
3.	.05 ×.3	.64 ×.9	3.15 ×.9	5.8 ×6.1	35.6 ×.42
4.	4 ×.02	5.3 ×.04	6.02 ×.04	7.81 ×15.2	.1628 ×100
5.	.9 ×.006	.67 ×.02	532 ×.07	3.86 ×4.04	418 ×.632

Perfect score: 25   My score: _____

PRE-TEST—Division

NAME _____

Chapter 8

Divide.

	a	b	c	d
1.	2)14.6	7)1.89	9).405	6).0114
2.	.3)6	.5)75	.02)42	.004)16
3.	.6).72	.3)6.3	.04).096	.003).015
4.	.04)3.2	.08)4.8	.002)7.26	.003)1.8
5.	.18)27	1.7).238	4.6)2.116	.38).3496

Perfect score: 20   My score: _____

Lesson 1 Division

> Place a decimal point in the quotient directly above the decimal point in the dividend. Then divide as if both numbers were whole numbers.

```
     17           1.7            .17           .017
6)102        6)10.2         6)1.02         6).102
  60           60             60             60
  ──           ──             ──             ──
  42           42             42             42
  42           42             42             42
  ──           ──             ──             ──
   0            0              0              0
```

Divide.

        a   *b*   *c*   *d*   *e*

1. 4)292   4)29.2   4)2.92   4).292   4).0292

2. 3)5.61   8).0216   7).231   4)4.64   6)25.2

3. 7)24.5   8).336   6).0162   4)24.4   3)1.68

Perfect score: 15   My score: _____

Problem Solving

Solve each problem.

1. A wire .8 inch long is to be cut into 4 pieces each the same length. How long will each piece be?

Each piece will be _____ inch long.

2. The same amount of flour was used in each of 3 batches of bread dough. A total of 6.9 kilograms was used. How much flour was in each batch?

_____ kilograms of flour was in each batch.

3. The combined thickness of 5 sheets of metal is .015 inch. Each sheet has the same thickness. How thick is each sheet?

Each sheet is _____ inch thick.

4. Each of 7 bolts has the same weight. Their total weight is .42 pound. How much does each bolt weigh?

Each bolt weighs _____ pound.

5. A machine can make 8 bolts in .008 hour. Each bolt takes the same amount of time. How long does it take to make 1 bolt?

It takes _____ hour to make 1 bolt.

6. Another machine takes 72.4 minutes to make 4 units. Each unit takes the same amount of time. How long does it take to make 1 unit?

It takes _____ minutes to make 1 unit.

7. A sheet of film is .072 centimeter thick. It is 6 times thicker than needed. What thickness is needed?

_____ centimeter is needed.

8. A sheet of film is .0672 inch thick. It is 8 times thicker than needed. What thickness is needed?

_____ inch is needed.

Perfect score: 8   My score: _____

Lesson 2 Division

Multiply the divisor and the dividend by 10, by 100, or by 1000 so the new divisor is a whole number.

```
                        40
.8)32  →  .8)32.0  →  8)320
                       320
         Multiply      ———
         by 10.          0


                        900
.05)45 → .05)45.00 →  5)4500
                      4500
         Multiply     ————
         by 100.         0
```

```
                            6500
.004)26 → .004)26.000 →  4)26000
                         24000
           Multiply      —————
           by 1000.       2000
                          2000
                         —————
                            0
```

Divide.

	a	b	c	d
1.	.4)7 2	.3)8 1	.7)3 5 7	.3)1 1 1
2.	.03)5 4	.04)9 6	.05)8 5	.08)2 9 6
3.	.002)6	.004)1 2	.006)2 4	.005)1 5 5

Perfect score: 12   My score: _____

Problem Solving

Solve each problem.

1. Dick put 72 kilograms of honey into jars. He put .4 kilogram into each jar. How many jars did he use?

   He used _____ jars.

2. A machine uses .3 gallon of fuel each hour. How many hours could the machine operate with 39 gallons of fuel?

   The machine could operate _____ hours.

3. Each metal bar weighs .04 kilogram. How many metal bars would weigh 520 kilograms?

   _____ metal bars would weigh 520 kilograms.

4. A machine uses .3 gallon of fuel each hour. At that rate, how many hours could the machine operate by using 195 gallons of fuel?

   The machine could operate _____ hours.

5. Eight-tenths gram of a chemical is put into each jar. How many jars can be filled with 192 grams of the chemical?

   _____ jars can be filled.

6. Each sheet of foil is .004 inch thick. How many sheets would be in a stack of foil that is 5 inches high?

   There would be _____ sheets.

7. How many nickels ($.05) are in $6?

   There are _____ nickels in $6.

8. How many nickels are in $10?

   There are _____ nickels in $10.

9. How many nickels are in $16?

   There are _____ nickels in $16.

1.	2.
3.	4.
5.	6.
7.	8.
9.	

Perfect score: 9   My score: _____

Lesson 3 Division

```
        23
.5)11.5 → .5)11.5 → 5)115
                    100
         Multiply    ‾‾‾
         by 10.       15
                      15
                      ‾‾
                       0
```

```
         7.1
.06).426 → .06).426 → 6)42.6
                     42 0
          Multiply   ‾‾‾‾
          by 100.      6
                       6
                       ‾
                       0
```

```
         700
.003)2.1 → .003)2.100 → 3)2100
                        2100
          Multiply      ‾‾‾‾
          by 1000.         0
```

Divide.

	a	b	c	d
1.	.4)7.2	.3).81	.8).392	.6)55.2
2.	.06).84	.04).068	.08).224	.07)2.52
3.	.002).008	.007).0042	.008).144	.009).0333
4.	.004).096	.09)6.3	.006).009	.7)8.4

Perfect score: 16   My score: _____

Lesson 4 Division

NAME _____

Divide.

	a	b	c	d
1.	.03)‾1.8	.06)‾28.8	.04)‾9.2	.05)‾1.5
2.	.003)‾2.4	.009)‾.45	.008)‾2.16	.005)‾2.5
3.	.02)‾7.4	.006)‾10.2	.08)‾27.2	.008)‾9.6
4.	.05)‾3.25	.06)‾4.2	.002)‾.4	.004)‾.56
5.	.3)‾75	.007)‾56.7	.09)‾74.7	.005)‾48

Perfect score: 20   My score: _____

Lesson 5   Division

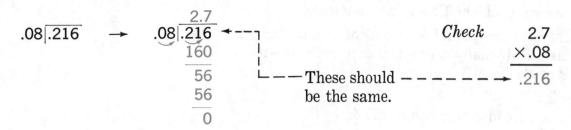

Divide. Check each answer.

	a	b	c
1.	3⟌1.44	6⟌17.4	5⟌.085
2.	.3⟌45	.003⟌12	.07⟌14
3.	.7⟌.98	.006⟌31.8	.08⟌.632
4.	.03⟌4.2	.004⟌7.2	.006⟌1.68
5.	.08⟌9.6	.003⟌84	6⟌9.6

Perfect score: 15   My score: _____

Problem Solving

Solve each problem. Check each answer.

1. Three-tenths pound of chemical is put into each container. How many containers can be filled with 5.4 pounds of chemical?

_____ containers can be filled.

2. Seventy-five hundredths pound of product Y is to be put into containers that hold .5 pound each. How many full containers will there be? What part of the next container will be filled?

There will be _____ full container.

_____ of the next container will be filled.

3. There are 10.2 pounds of ball bearings in a box. Each bearing weighs .006 pound. How many bearings are in the box?

There are _____ bearings in the box.

4. A machine processes 1.95 pounds of chemical every 3 hours. At that rate, how many pounds of chemical are processed in 1 hour?

_____ pounds are processed.

5. Each sheet of metal is .005 inch thick. How many sheets of metal would there be in a stack that is 4.2 inches high?

There would be _____ sheets.

6. A wire that is .6 meter long is cut into pieces of the same length. Each piece is .06 meter long. How many pieces of wire are there?

There are _____ pieces.

7. Suppose in problem 6 each piece of wire is .006 meter long. How many pieces are there?

There are _____ pieces.

Perfect score: 8   My score: _____

Lesson 6 Division

$$.25\overline{)1} \longrightarrow .25\overline{)100} \longrightarrow 25\overline{)100}$$

with quotient 4, 100, 0

$$2.7\overline{)3.78} \longrightarrow 2.7\overline{)3.78} \longrightarrow 27\overline{)37.8}$$

with quotient 1.4, 270, 108, 108, 0

Divide.

	a	b	c	d
1.	$.20\overline{)1}$	$.15\overline{)9}$	$.028\overline{)1\,4}$	$.012\overline{)6}$
2.	$1.2\overline{)3.9\,6}$	$.42\overline{).7\,5\,6}$	$.18\overline{).8\,2\,8}$	$2.5\overline{).6\,2\,5}$
3.	$.67\overline{).3\,8\,8\,6}$	$.45\overline{)1.2\,1\,5}$	$7.3\overline{)3\,0.6\,6}$	$4.3\overline{).1\,3\,7\,6}$
4.	$.025\overline{)7\,5}$	$.36\overline{)1.5\,1\,2}$	$5.4\overline{).3\,9\,4\,2}$	$.53\overline{).6\,3\,6}$

Perfect score: 16   My score: _____

Problem Solving

Solve each problem.

1. A carton of items weighs 28.8 pounds. Each item weighs 3.6 pounds. How many items are in the carton?

_____ items are in the carton.

2. There is .444 pound of chemical to be put into tubes. Each tube holds .12 pound. How many tubes will be filled? How much of another tube is filled?

_____ tubes will be filled.

_____ of the next tube will be filled.

3. A stack of cardboard is 52 inches high. Each piece is .65 inch thick. How many pieces of cardboard are in the stack?

_____ pieces are in the stack.

4. How many pieces each .25 inch long can be cut from a wire that is 2 inches long?

_____ pieces can be cut from the wire.

5. How many pieces each .5 inch long can be cut from the wire described in problem 4?

_____ pieces can be cut from the wire.

6. Each bag of flour weighs 2.2 pounds. How many such bags can be filled by using 11 pounds of flour?

_____ bags can be filled.

7. Consider the numbers named by .016; 1.6; and .16. What is the quotient if you divide the greatest number by the least number?

The quotient is _____.

8. Suppose in problem 7 you divide the least number by the greatest number. What is the quotient?

The quotient is _____.

1.	2.
3.	4.
5.	6.
7.	8.

Perfect score: 9   My score: _____

Lesson 7 Division

NAME _____

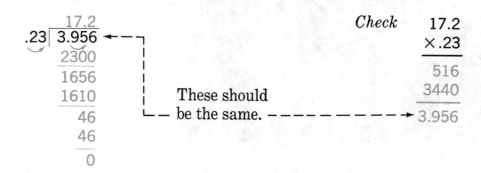

Divide. Check each answer.

          a   *b*   *c*

1.   .73) 5.986   5.6) .672   .15) 75

2.   2.1) 6.93   .15) 18   .083) 6.308

3.   .37) .1739   1.6) 4.48   .53) 4.876

Perfect score: 9   My score: _____

Problem Solving

Solve each problem.

1. A 10-ounce can of fruit costs $.59. Find the cost per ounce.

The cost per ounce is $ _____ .

2. A 4.75-ounce bar of hand soap costs $.57. Find the cost per ounce.

The cost per ounce is $ _____ .

3. A 6.5-ounce can of tuna costs $.91. Find the cost per ounce.

The cost per ounce is $ _____ .

4. A 6.4-ounce tube of toothpaste costs $2.24. Find the cost per ounce.

The cost per ounce is $ _____ .

5. A 147-ounce box of detergent costs $5.88. Find the cost per ounce.

The cost per ounce is $ _____ .

6. Each fish stick weighs .8 ounce. One fish stick costs $.08. Find the cost per ounce.

The cost per ounce is $ _____ .

7. Each slice of cheese weighs .75 ounce. There are 20 slices of cheese in a package. The package costs $2.40. Find the cost per ounce.

The cost per ounce is $ _____ .

Perfect score: 7   My score: _____

CHAPTER 8 TEST

NAME _____

Divide.

	a	b	c	d
1.	8).184	3).42	4)14.8	6).0306
2.	.05)55	.003)36	.7)42	.04)84
3.	.4)9.2	.6).84	.03).072	.004).028
4.	.006)5.4	.07)4.9	.007).63	.004)41.2
5.	.36)9	3.8)5.32	.42)1.092	4.5).3285

Perfect score: 20   My score: _____

PRE-TEST—Metric Measurement

NAME _____

Chapter 9

	a	b
1.	7 meters = _____ millimeters	85 millimeters = _____ meter
2.	1.9 meters = _____ centimeters	4.5 centimeters = _____ meter
3.	6 kilometers = _____ meters	7 meters = _____ kilometer
4.	5 liters = _____ milliliters	4000 milliliters = _____ liters
5.	3.5 kiloliters = _____ liters	6.5 liters = _____ kiloliter
6.	8.2 kilograms = _____ grams	255 grams = _____ kilogram

Find the area of each rectangle.

a

b

7.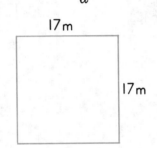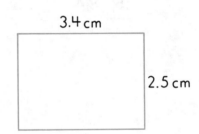

_____ square meters   _____ square centimeters

Find the volume of each rectangular solid.

a

b

8.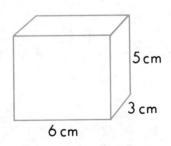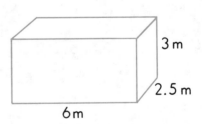

_____ cubic centimeters   _____ cubic meters

Perfect score: 16   My score: _____

Lesson 1 Length

NAME _____

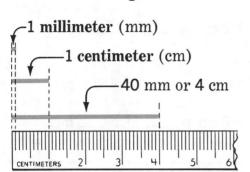

1 millimeter (mm)
1 centimeter (cm)
40 mm or 4 cm

10 mm = 1 cm	1 mm = .1 cm
1000 mm = 1 m	1 mm = .001 m
100 cm = 1 m	1 cm = .01 m
1000 m = 1 km	1 m = .001 km

100 cm or 1 **meter** (m)

A distance of 1000 meters is 1 **kilometer** (km).

Measure the following to the nearest meter.

                a   *b*

1. height of classroom door _____ m   width of room _____ m

2. width of classroom door _____ m   length of room _____ m

3. One kilometer is about the length of 5 blocks. About how many kilometers do you live from school? _____ km

Find the length of each line segment to the nearest centimeter.

4. _____ cm

5. _____ cm

Find the length of each line segment to the nearest millimeter.

6. _____ mm

7. _____ mm

Draw a line segment for each measurement.

8. 6 cm

9. 8 cm

10. 35 mm

Perfect score: 12   My score: _____

Lesson 2 Units of Length

NAME _____

160 m = __?__ km	7.2 cm = __?__ mm
1 m = .001 km	1 cm = 10 mm
(160 × 1) m = (160 × .001) km	(7.2 × 1) cm = (7.2 × 10) mm
160 m = __.16__ km	7.2 cm = __72__ mm

Find the length of each line segment to the nearest millimeter.
Then give the length in centimeters and in meters.

	a	b	c
1.	_____ mm	_____ cm	_____ m
2.	_____ mm	_____ cm	_____ m
3.	_____ mm	_____ cm	_____ m
4.	_____ mm	_____ cm	_____ m
5.	_____ mm	_____ cm	_____ m

Complete the following.

	a	b	c
6.	54 mm = _____ cm	8 m = _____ mm	234 m = _____ km
7.	1.6 m = _____ cm	.9 cm = _____ mm	58 mm = _____ cm
8.	612 mm = _____ m	4 km = _____ m	13 mm = _____ cm
9.	.02 m = _____ mm	.75 km = _____ m	34.5 m = _____ km
10.	707 cm = _____ m	.005 m = _____ cm	465 mm = _____ cm

Perfect score: 30   My score: _____

Lesson 3 Area

NAME _____

To find the *area measure* (A) of a rectangle, multiply the measure of its *length* (l) by the measure of its *width* (w).

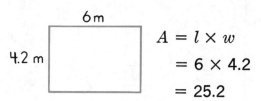

$A = l \times w$
$= 6 \times 4.2$
$= 25.2$

The area is __25.2__ square meters.

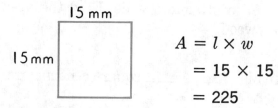

$A = l \times w$
$= 15 \times 15$
$= 225$

The area is _____ square millimeters.

Measure the sides as indicated. Then find the area of each rectangle.

          a   *b*

1. _____ cm _____ cm
    _____ cm _____ cm

    _____ square centimeters   _____ square centimeters

2. _____ mm _____ mm
    _____ mm _____ mm

    _____ square millimeters   _____ square millimeters

Find the area of each rectangle described below.

	length	width	area
3.	5.8 km	3 km	_____ square kilometers
4.	4.5 m	4.5 m	_____ square meters
5.	32.4 m	16 m	_____ square meters

Perfect score: 15   My score: _____

Problem Solving

Solve each problem.

1. A rectangular piece of plywood is 150 centimeters long. It is 75 centimeters wide. Find the area of the piece of plywood.

   The area is _____ square centimeters.

2. A rectangular piece of carpet is 6 meters long and 4.5 meters wide. How many square meters of carpet is that?

   That is _____ square meters of carpet.

3. A rectangular floor is 8.5 meters long and 6.5 meters wide. How many square meters of floor are there?

   There are _____ square meters of floor.

4. A playground is shaped like a rectangle. Its length is 140 meters. Its width is 60 meters. Find the area of the playground.

   The area is _____ square meters.

5. A rectangular poster is 35 centimeters wide and 52 centimeters long. Find the area of the poster.

   The area is _____ square centimeters.

6. A rectangular postcard is 8.5 centimeters wide and 14 centimeters long. Find the area of the postcard.

   The area is _____ square centimeters.

7. A rectangular field is 1000 meters long and 800 meters wide. How many square meters are in that field?

   There are _____ square meters in the field.

Perfect score: 7   My score: _____

Lesson 4 Volume

NAME _____

To determine the *volume measure* (V) of a rectangular solid, find the product of the measure of its *length* (l), the measure of its *width* (w), and the measure of its *height* (h).

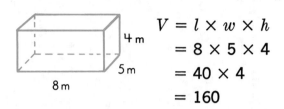

$V = l \times w \times h$
$= 8 \times 5 \times 4$
$= 40 \times 4$
$= 160$

The volume is __160__ cubic meters.

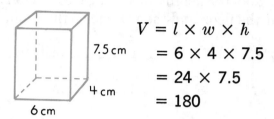

$V = l \times w \times h$
$= 6 \times 4 \times 7.5$
$= 24 \times 7.5$
$= 180$

The volume is _____ cubic centimeters.

Find the volume of each rectangular solid below.

        a   *b*   *c*

1.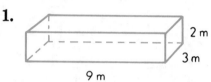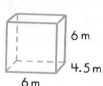

_____ cubic meters   _____ cubic meters   _____ cubic meters

2.

_____ cubic centimeters   _____ cubic millimeters   _____ cubic meter

Find the volume of each rectangular solid described below.

	length	width	height	volume
3.	7 m	6 m	5 m	_____ cubic meters
4.	9.2 cm	4.5 cm	3 cm	_____ cubic centimeters
5.	8.5 mm	8.5 mm	8.5 mm	_____ cubic millimeters
6.	7.2 cm	6.2 cm	5.2 cm	_____ cubic centimeters

Perfect score: 10   My score: _____

Problem Solving

Solve each problem.

1. Find a rectangular room. Measure its length, its width, and its height to the nearest meter. Find the area of the floor and the volume of the room.

    length: _____ meters

    width: _____ meters

    height: _____ meters

    floor area: _____ square meters

    volume: _____ cubic meters

2. Find a rectangular box. Measure its length, its width, and its height to the nearest centimeter. Find the area of the box top and the volume of the box.

    length: _____ centimeters

    width: _____ centimeters

    height: _____ centimeters

    top area: _____ square centimeters

    volume: _____ cubic centimeters

3. A shipping crate is 1.2 meters high, .8 meter wide, and 2.5 meters long. Find the volume of the crate.

    The volume is _____ cubic meters.

4. A book is 28 centimeters long, 21 centimeters wide, and 1 centimeter thick. How much space does the book occupy?

    It occupies _____ cubic centimeters of space.

5. A hole was dug 12.5 meters long, 10.5 meters wide, and 2 meters deep. How many cubic meters of earth were removed?

    _____ cubic meters of earth were removed.

Perfect score: 13   My score: _____

Lesson 5 Capacity

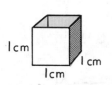

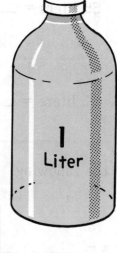

A box this size can hold **1 milliliter**.

A teaspoon can hold about 5 milliliters.

A 1-**liter** bottle can hold 1000 milliliters.

1 liter = 1000 milliliters (ml)
1000 liters = 1 kiloliter (kl)

1 ml = .001 liter
1 liter = .001 kl

Solve each problem.

1. Suppose you drank 5 liters of milk each week. How many milliliters would that be?

That would be _____ milliliters of milk.

2. Each can holds 4 liters of paint. How many liters of paint are in 6 such cans?

_____ liters are in six cans.

3. Each bottle holds 40 milliliters of vanilla. How many such bottles can be filled from 1 liter of vanilla?

_____ bottles can be filled.

4. There are 8000 liters of water in a pool. How many kiloliters of water are in that pool?

_____ kiloliters of water are in that pool.

5. A ship is carrying 480 barrels of oil. Each barrel contains 200 liters of oil. How many liters of oil is the ship carrying?

The ship is carrying _____ liters of oil.

Perfect score: 5   My score: _____

113

Lesson 6 Units of Capacity

65 liters = _____?_____ ml

1 liter = 1000 ml
(65 × 1) liters = (65 × 1000) ml

65 liters = __65,000__ ml

5.2 liters = _____?_____ kl

1 liter = .001 kl
(5.2 × 1) liters = (5.2 × .001) kl

5.2 liters = __.0052__ kl

Complete the following.

	a	b
1.	7 liters = _____ ml	.5 liter = _____ ml
2.	5 ml = _____ liter	4500 ml = _____ liters
3.	7.5 kl = _____ liters	2.54 kl = _____ liters
4.	600 liters = _____ kl	7.5 liters = _____ kl
5.	3.4 liters = _____ ml	300 ml = _____ liter
6.	3 kl = _____ liters	300 liters = _____ kl
7.	.6 liter = _____ kl	24 kl = _____ liters
8.	47 ml = _____ liter	.75 liter = _____ ml

Solve.

9. Consider filling the tank shown with water. How many milliliters would the tank hold? How many liters? (Hint: A 1 cubic-centimeter container can hold 1 milliliter of water.)

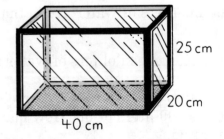

The tank would hold _____ milliliters.

The tank would hold _____ liters.

Perfect score: 18   My score: _____

Lesson 7   Weight

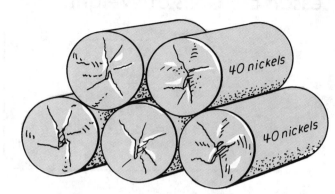

2 paper clips weigh about 1 **gram** (g).

1 nickel weighs about 5 grams.

200 nickels weigh about 1000 grams or 1 **kilogram** (kg).

1 g = 1000 milligrams (mg)
1 kg = 1000 g

1 mg = .001 g
1 g = .001 kg

Solve each problem.

1. There are 100 paper clips in a box. How many grams would a box of paper clips weigh?

It would weigh _____ grams.

2. There are 20 boxes of paper clips in a carton. How many grams would a carton of paper clips weigh?

The carton would weigh _____ grams.

3. How many grams would a roll of 40 nickels weigh?

The roll would weigh _____ grams.

4. How many nickels would weigh 5 kilograms?

_____ nickels would weigh 5 kilograms.

5. A truck is hauling 6 crates. Each crate weighs 35 kilograms. How much do all the crates weigh?

They weigh _____ kilograms.

1.	2.
3.	4.
5.	

Perfect score: 5   My score: _____

Lesson 8 Units of Weight

73 kg = ___?___ g

1 kg = 1000 g
(73 × 1) kg = (73 × 1000) g

73 kg = __73,000__ g

54 mg = ___?___ g

1 mg = .001 g
(54 × 1) mg = (54 × .001) g

54 mg = __.054__ g

Complete the following.

a   *b*

1. 8 kg = _____ g   7.5 kg = _____ g

2. 4500 mg = _____ g   38 g = _____ kg

3. 6 g = _____ mg   640 mg = _____ g

4. .05 kg = _____ g   4.5 kg = _____ g

5. .007 kg = _____ g   7000 g = _____ kg

6. 5.5 g = _____ mg   .21 g = _____ mg

7. .4 kg = _____ g   345 g = _____ kg

8. 607 mg = _____ g   8.9 g = _____ mg

9. 52 g = _____ mg   975 g = _____ kg

Solve.

10. Consider this tank filled with water. Assume that 1 liter of water weighs 1 kilogram. How many kilograms would the water in the tank weigh?

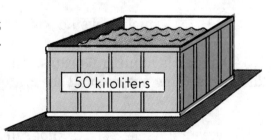

The water would weigh _____ kilograms.

Perfect score: 19   My score: _____

CHAPTER 9 TEST

NAME _____

Complete the following.

                a   b

1.   9 m = _____ mm   365 mm = _____ m

2.   7.2 m = _____ cm   8.4 cm = _____ m

3.   26 km = _____ m   17 m = _____ km

4.   4.51 liters = _____ ml   65 ml = _____ liter

5.   6.8 kl = _____ liters   3500 liters = _____ kl

6.   .9 kg = _____ g   785 g = _____ kg

7.   2 g = _____ mg   .998 kg = _____ g

8.   1500 m = _____ km   .45 kg = _____ g

Find the area of each rectangle.

                a   b

9. 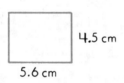   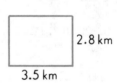

    _____ square centimeters   _____ square kilometers

Find the volume of each rectangular solid.

                a   b

10.   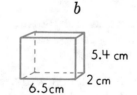

    _____ cubic meters   _____ cubic centimeters

Perfect score: 20   My score: _____

PRE-TEST—Measurement Chapter 10

Complete the following.

                   a   b

1. 3 feet = _____ inches   48 inches = _____ feet

2. 6 yards = _____ feet   24 feet = _____ yards

3. 2 miles = _____ feet   3 miles = _____ yards

4. 3 pints = _____ cups   14 cups = _____ pints

5. 7 quarts = _____ pints   8 quarts = _____ gallons

6. 2 pounds = _____ ounces   2000 pounds = _____ ton

7. 6 minutes = _____ seconds   180 seconds = _____ minutes

8. 48 hours = _____ days   120 minutes = _____ hours

9. 4 feet 8 inches = _____ inches

10. 3 gallons 3 quarts = _____ quarts

11. 2 pounds 6 ounces = _____ ounces

12. 2 minutes 30 seconds = _____ seconds

Find the area of each figure.

                 a   b

13. 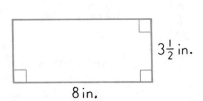   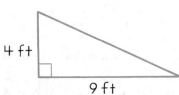

      _____ square inches   _____ square feet

Solve.

14. A shoe box is 6 inches wide, 11 inches long, and 5 inches high. Find the volume of the box.

The volume is _____ cubic inches.

Perfect score: 23   My score: _____

Lesson 1 Length

1 foot (ft) = 12 inches (in.)	1 in. = $\frac{1}{12}$ ft
1 yard (yd) = 3 ft	1 ft = $\frac{1}{3}$ yd
1 yd = 36 in.	1 in. = $\frac{1}{36}$ yd
1 mile (mi) = 5280 ft	1 mi = 1760 yd

36 in. = _____?_____ ft

1 in. = $\frac{1}{12}$ ft
(36 × 1) in. = (36 × $\frac{1}{12}$) ft

36 in. = ___3___ ft

6 ft 4 in. = _____?_____ in.

1 ft = 12 in.
6 ft = (6 × 12) or 72 in.
6 ft 4 in. = (72 + 4) in.

6 ft 4 in. = ___76___ in.

Complete the following.

    a   *b*

1. 6 ft = _____ in.   60 in. = _____ ft

2. 9 yd = _____ ft   12 ft = _____ yd

3. 5 yd = _____ in.   144 in. = _____ yd

4. 3 mi = _____ ft   3 mi = _____ yd

5. 5 yd = _____ ft   18 in. = _____ ft

6. 2 mi = _____ ft   5 mi = _____ yd

7. 5 ft 4 in. = _____ in.

8. 3 yd 5 in. = _____ in.

9. 5 yd 2 ft = _____ ft

10. 9 ft 6 in. = _____ in.

11. 1 mi 750 ft = _____ ft

Perfect score: 17 My score: _____

Problem Solving

Solve each problem.

1. The top of a doorway is 84 inches above the floor. What is the height of the doorway in feet?

The doorway is _____ feet high.

2. The distance along the foul line from home plate to the right field fence is 336 feet. What is this distance in yards?

This distance is _____ yards.

3. A kite string is 125 yards long. How many feet long is the string?

The string is _____ feet long.

4. A snake is 4 feet 3 inches long. What is the length of the snake in inches?

Its length is _____ inches.

5. A rope is 5 feet 9 inches long. What is the length of the rope in inches?

The rope is _____ inches long.

6. The distance across a street is 15 yards 1 foot. What is this distance in feet?

This distance is _____ feet.

7. A doorway is 2 feet 8 inches wide. What is the width of the doorway in inches?

The doorway is _____ inches wide.

8. Helen is 4 feet 11 inches tall. What is her height in inches?

Helen is _____ inches tall.

9. Jerry is 5 feet 4 inches tall. What is his height in inches?

Jerry is _____ inches tall.

Perfect score: 9   My score: _____

Lesson 2 Area

To determine the *area measure* (A) of a right triangle, find *one half* the product of the measure of its *base* (b) and the measure of its *height* (h).

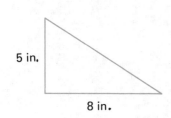

$A = \frac{1}{2} \times b \times h$
$= \frac{1}{2} \times (8 \times 5)$
$= \frac{1}{2} \times 40$
$= 20$

The area is __20__ square inches.

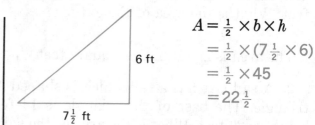

$A = \frac{1}{2} \times b \times h$
$= \frac{1}{2} \times (7\frac{1}{2} \times 6)$
$= \frac{1}{2} \times 45$
$= 22\frac{1}{2}$

The area is _____ square feet.

Find the area of each right triangle below.

            a   *b*   *c*

1.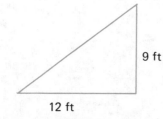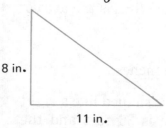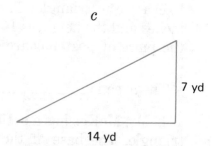

_____ square feet   _____ square inches   _____ square yards

2.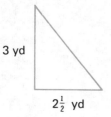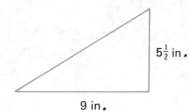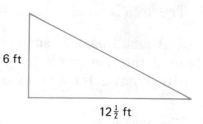

_____ square yards   _____ square inches   _____ square feet

Find the area of each right triangle described below.

	base	height	area
3.	8 ft	9 ft	_____ square feet
4.	7 yd	5 yd	_____ square yards
5.	$4\frac{1}{2}$ in.	6 in.	_____ square inches
6.	5 ft	$3\frac{1}{2}$ ft	_____ square feet
7.	$3\frac{3}{4}$ in.	2 in.	_____ square inches

Perfect score: 11   My score: _____

Problem Solving

Solve each problem.

1. The edges of a flower garden form a right triangle. The base of the triangle is 16 feet and the height is 8 feet. What is the area of the garden?

    The area is _____ square feet.

2. A sailboat has a sail which is shaped like a right triangle. The base of the triangle is 14 feet and the height is 20 feet. What is the area of the sail?

    The area is _____ square feet.

3. Anne has a piece of poster board which is shaped like a right triangle. The base of the triangle is 28 inches and the height is $16\frac{1}{2}$ inches. What is the area of the piece of poster board?

    The area is _____ square inches.

4. Mr. McKee has a patio which is shaped like a right triangle. The base of the triangle is 36 feet and the height is 12 feet. What is the area of the patio?

    The area is _____ square feet.

5. A small park is shaped like a right triangle. The base of the triangle is 160 yards and the height is 120 yards. What is the area of the park?

    The area is _____ square yards.

6. Nelson has a piece of sheet metal which is shaped like a right triangle. The base of the triangle is 16 inches and the height is $12\frac{1}{2}$ inches. What is the area of the piece of sheet metal?

    The area is _____ square inches.

7. Mrs. Jones has a piece of material which is shaped like a right triangle. The base of the triangle is $25\frac{1}{2}$ inches and the height is 18 inches. What is the area of the piece of material?

    The area is _____ square inches.

Perfect score: 7   My score: _____

NAME _____

Lesson 3   Area and Volume

Find the area of each right triangle or rectangle below.

                    a   *b*   *c*

1.
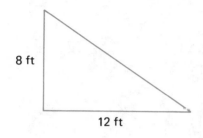 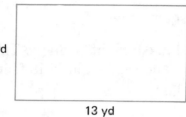 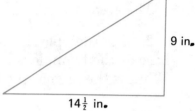

_____ square feet   _____ square yards   _____ square inches

2.
 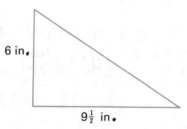 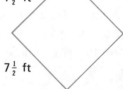

_____ square yards   _____ square inches   _____ square feet

Find the volume of each rectangular solid described below.

	length	width	height	volume
3.	7 yd	5 yd	3 yd	_____ cubic yards
4.	9 in.	5 in.	4½ in.	_____ cubic inches
5.	6 ft	3¼ ft	9 ft	_____ cubic feet
6.	5½ yd	3 yd	7 yd	_____ cubic yards
7.	3¼ in.	2¾ in.	4 in.	_____ cubic inches
8.	6½ ft	5 ft	4½ ft	_____ cubic feet
9.	3 in.	5¼ in.	3½ in.	_____ cubic inches
10.	9¼ ft	8¾ ft	5 ft	_____ cubic feet

Perfect score: 14   My score: _____

Problem Solving

Solve each problem.

1. A basketball court is shaped like a rectangle. The length is 84 feet and the width is 50 feet. What is the area of the court?

    The area is _____ square feet.

2. A garden plot is shaped like a right triangle. The base of the triangle is 50 feet and the height is 18 feet. What is the area of the triangle?

    The area is _____ square feet.

3. A suitcase is 32 inches long, 16 inches wide, and 6 inches deep. What is the volume of the suitcase?

    The volume is _____ cubic inches.

4. Mrs. Langley has a flower bed which is shaped like a right triangle. The base of the triangle is $12\frac{1}{2}$ feet and the height is 6 feet. What is the area of the flower bed?

    The area is _____ square feet.

5. A plot of land is shaped like a rectangle. It is 280 yards long and 90 yards wide. What is the area of the plot?

    The area is _____ square yards.

6. A box is 9 inches long, $6\frac{1}{2}$ inches wide, and $1\frac{1}{2}$ inches deep. What is the volume of the box?

    The volume is _____ cubic inches.

7. A rectangular tabletop is 72 inches long and 36 inches wide. What is the area of the tabletop?

    The area is _____ square inches.

8. A brick is 8 inches long, 3 inches wide, and 2 inches high. How much space does the brick occupy?

    The brick occupies _____ cubic inches of space.

Perfect score: 8   My score: _____

Lesson 4 Capacity

> 1 pint (pt) = 2 cups
> 1 quart (qt) = 2 pt
> 1 gallon (gal) = 4 qt

> 1 cup = $\frac{1}{2}$ pt
> 1 pt = $\frac{1}{2}$ qt
> 1 qt = $\frac{1}{4}$ gal

5 pt = ____?____ qt

1 pt = $\frac{1}{2}$ qt
5 pt = ($\frac{1}{2}$ × 5) qt
5 pt = ____$2\frac{1}{2}$____ qt

3 gal 2 qt = ____?____ qt

1 gal = 4 qt
3 gal = (3 × 4) or 12 qt
3 gal 2 qt = (12 + 2) qt

3 gal 2 qt = ____14____ qt

Complete the following.

	a	b
1.	3 pt = ____ cups	8 cups = ____ pt
2.	5 qt = ____ pt	10 pt = ____ qt
3.	4 gal = ____ qt	11 qt = ____ gal
4.	24 qt = ____ gal	15 pt = ____ qt
5.	2 pt 1 cup = ____ cups	
6.	5 gal 3 qt = ____ qt	
7.	2 qt 1 pt = ____ pt	
8.	4 gal 3 qt = ____ qt	

9. An aquarium holds 3 gallons 3 quarts of water. How many quarts would this be? How many pints? How many cups?

This would be _____ quarts.

This would be _____ pints.

This would be _____ cups.

Perfect score: 15   My score: _____

Lesson 5 Weight and Time

NAME _____

| 1 pound (lb) = 16 ounces (oz) | 1 oz = $\frac{1}{16}$ lb |
| 1 ton = 2000 lb | |

1 minute (min) = 60 seconds (sec)	1 sec = $\frac{1}{60}$ min
1 hour = 60 min	1 min = $\frac{1}{60}$ hour
1 day = 24 hours	1 hour = $\frac{1}{24}$ day

80 oz = ____?____ lb   1 min 12 sec = ____?____ sec

1 oz = $\frac{1}{16}$ lb   1 min = 60 sec
80 oz = (80 × $\frac{1}{16}$) lb   1 min 12 sec = (60 + 12) sec

80 oz = ____5____ lb   1 min 12 sec = ____72____ sec

Complete the following.

                  a   *b*

1. 72 lb = _____ oz   80 oz = _____ lb

2. 4 tons = _____ lb   6000 lb = _____ tons

3. 3 min = _____ sec   120 sec = _____ min

4. 5 hours = _____ min   360 min = _____ hours

5. 5 days = _____ hours   144 hours = _____ days

6. 3 lb 12 oz = _____ oz   5 lb 6 oz = _____ oz

7. 3 tons 500 lb = _____ lb

8. 2 hr 45 min = _____ min

9. 4 days 12 hours = _____ hours

10. 4 hours 20 min = _____ min

11. 2 days 8 hours = _____ hours

Perfect score: 17   My score: _____

CHAPTER 10 TEST

NAME _____

Complete the following.

                a   *b*

1. 8 ft = _____ in.   72 in. = _____ ft

2. 5 yd = _____ ft   30 ft = _____ yd

3. 2 mi = _____ ft   1 mi = _____ yd

4. 6 pt = _____ cups   11 cups = _____ pt

5. 18 qt = _____ pt   12 pt = _____ qt

6. 5 gal = _____ qt   4000 lb = _____ tons

7. 6 lb = _____ oz   32 oz = _____ lb

8. 5 min = _____ sec   120 sec = _____ min

9. 5 ft 10 in. = _____ in.   2 days = _____ hours

10. 4 gal 1 qt = _____ qt   72 hours = _____ days

11. 4 lb 12 oz = _____ oz

12. 3 hours 30 min = _____ min

Solve.

13. A box is 36 inches long, 12 inches wide, and 10 inches high. Find the volume of the box.

The volume of the box is _____ cubic inches.

Find the area of each figure.

                a   *b*

14. 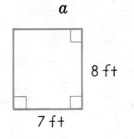   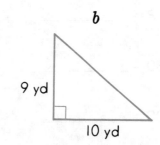

_____ square feet   _____ square yards

Perfect score: 25   My score: _____

PRE-TEST—Percent

Chapter 11

Complete the following.

　　　　　a　　　　　　　　　　　　　　*b*

1. $\frac{7}{100}$ = _____ %　　　　$\frac{9}{10}$ = _____ %

2. $\frac{7}{20}$ = _____ %　　　　$\frac{13}{25}$ = _____ %

3. .07 = _____ %　　　　.4 = _____ %

4. .135 = _____ %　　　　1.35 = _____ %

Change each percent to a decimal.

　　　　　a　　　　　　　　　　　　　　*b*

5. 6% = _____　　　　67% = _____

6. 6.25% = _____　　　　125% = _____

Change each percent to a fraction in simplest form.

　　　　　a　　　　　　　　　　　　　　*b*

7. 9% = _____　　　　20% = _____

8. 45% = _____　　　　56% = _____

Complete the following.

　　　　　a　　　　　　　　　　　　　　*b*

9. 1% of 59 = _____　　　　10% of 75 = _____

10. 100% of 84 = _____　　　　45% of 89 = _____

11. 75% of 48 = _____　　　　83% of 147 = _____

12. 12% of 180 = _____　　　　7.5% of 840 = _____

13. 6.75% of 500 = _____　　　　8.5% of 96.4 = _____

14. 12.5% of 420 = _____　　　　7% of 79.3 = _____

Perfect score: 28　　My score: _____

NAME _____

Lesson 1   Percent

The symbol % (read **percent**) means $\frac{1}{100}$ or .01.

$3\% = 3 \times \frac{1}{100}$ or $3\% = 3 \times .01$ | $17\% = 17 \times \frac{1}{100}$ or $17\% = 17 \times .01$

= $\frac{3}{100}$ = .03 | = _____   = _____

Complete the following.

	percent	fraction	decimal
1.	1%		
2.	7%		
3.	29%		
4.	47%		
5.	53%		
6.	21%		
7.	83%		
8.	49%		
9.	61%		
10.	9%		
11.	37%		
12.	77%		
13.	91%		
14.	33%		

Perfect score: 28   My score: _____

Lesson 2 Percent and Fractions

NAME _____

Study how a percent is changed to a fraction or mixed numeral in simplest form.

$75\% = 75 \times \frac{1}{100}$ $125\% = 125 \times \frac{1}{100}$
$\quad = \frac{75}{100}$ $\quad = \frac{125}{100}$
$\quad = \underline{\quad \frac{3}{4} \quad}$ $\quad = \frac{5}{4}$ or _____

Study how a fraction or mixed numeral is changed to a percent.

$\frac{1}{2} = \frac{1}{2} \times \frac{50}{50}$ $1\frac{3}{4} = \frac{7}{4} \times \frac{25}{25}$
$\quad = \frac{50}{100}$ $\quad = \frac{175}{100}$
$\quad = 50 \times \frac{1}{100}$ $\quad = 175 \times \frac{1}{100}$
$\quad = \underline{\ 50\% \ }$ $\quad = $ _____ %

Change each of the following to a fraction or mixed numeral in simplest form.

	a	b	c
1.	25% = _____	45% = _____	160% = _____
2.	65% = _____	120% = _____	24% = _____
3.	78% = _____	55% = _____	260% = _____
4.	70% = _____	144% = _____	86% = _____
5.	95% = _____	40% = _____	180% = _____

Change each of the following to a percent.

	a	b	c
6.	$\frac{1}{5}$ = _____	$\frac{3}{4}$ = _____	$\frac{1}{20}$ = _____
7.	$2\frac{7}{50}$ = _____	$\frac{3}{5}$ = _____	$1\frac{1}{5}$ = _____
8.	$\frac{9}{10}$ = _____	$\frac{7}{25}$ = _____	$2\frac{1}{4}$ = _____
9.	$1\frac{3}{5}$ = _____	$\frac{3}{10}$ = _____	$\frac{4}{25}$ = _____
10.	$\frac{7}{20}$ = _____	$\frac{31}{50}$ = _____	$1\frac{2}{5}$ = _____

Perfect score: 30   My score: _____

NAME _____

Lesson 3 Percent and Decimals

Study how a percent is changed to a decimal.

12.5% = 12.5 × .01
= __.125__

1.25% = 1.25 × .01
= _____

Study how a decimal is changed to a percent.

.7 = .70
= 70 × .01
= __70%__

.245 = 24.5 × .01
= _____ %

Change each of the following to a decimal.

	a	b	c
1.	13.5% =	37% =	6.25% =
2.	6.5% =	4.75% =	2.75% =
3.	7% =	62.5% =	8.5% =
4.	32.5% =	8.75% =	9.5% =
5.	8.25% =	17.5% =	3.75% =
6.	.75% =	7.25% =	1.75% =

Change each of the following to a percent.

	a	b	c
7.	.6 =	.52 =	.325 =
8.	.2475 =	.8 =	.65 =
9.	.145 =	.1675 =	.5 =
10.	.06 =	.007 =	.0625 =
11.	.075 =	.0075 =	.005 =
12.	.9 =	.19 =	.385 =

Perfect score: 36   My score: _____

Problem Solving

Solve each problem.

1. Three fourths of the pupils in class are girls. What percent of the pupils are girls?

   _____ of the pupils are girls.

2. Mr. Beck received 65% of the votes cast. What fractional part of the votes did he receive?

   He received _____ of the votes.

3. Marty made a base hit on 25% of his official times at bat. What is his batting average? (Note: Batting averages are usually expressed as thousandths.)

   His average is _____.

4. Four fifths of the pupils are in the gym. What percent of the pupils are in the gym?

   _____ of the pupils are in the gym.

5. A farmer has 45% of a field plowed. Write a fraction to tell how much of the field is plowed.

   _____ of the field is plowed.

6. The Cubs won 61.5% of their games last year. How can this percent be expressed as a decimal?

   61.5% can be expressed as _____.

7. A certain player has a fielding average of .987. How can this fielding average be expressed as a percent?

   This average can be expressed as _____.

8. Seven tenths of the customers at the Caribbean Market come in the morning. What percent of the customers come in the morning?

   _____ of the customers come in the morning.

Perfect score: 8   My score: _____

Lesson 4   Percent of a Number

Study how fractions are used to find a percent of a number.

75% of 60 = 75% × 60
$= \frac{75}{100} \times 60$
$= \frac{3}{4} \times \frac{60}{1}$
$= \frac{3 \times 60}{4 \times 1}$
$= \frac{180}{4}$ or 45

125% of 37.5 = 125% × 37.5
$= \frac{125}{100} \times 37.5$
$= \frac{5}{4} \times \frac{375}{10}$
$= \frac{5 \times 375}{4 \times 10}$
$= \frac{1875}{40}$ or $46 \frac{7}{8}$

75% of 60 = __45__

125% of 37.5 = _____

Write each answer in simplest form.

           a   *b*

1.   6% of 75 = _____   108% of 63.5 = _____

2.   50% of 32 = _____   75% of 12.6 = _____

3.   20% of 68 = _____   25% of 72.8 = _____

4.   5% of 48 = _____   15% of 52.4 = _____

5.   104% of 35 = _____   136% of 7.5 = _____

6.   55% of 5.25 = _____   80% of 160 = _____

7.   166% of 60 = _____   90% of 1.8 = _____

8.   72% of 7.25 = _____   140% of 240 = _____

9.   60% of 9.8 = _____   250% of 90 = _____

10.   100% of 725 = _____   40% of 9.6 = _____

Perfect score: 20   My score: _____

Problem Solving

Solve each problem.

1. Twenty-five percent of the workers are on the third shift. There are 132 workers in all. How many of them are on the third shift?

   _____ are on the third shift.

2. The enrollment at Franklin School has increased 20% from last year. The enrollment last year was 750. By how many pupils has the enrollment increased?

   The enrollment has increased by _____ pupils.

3. Ms. Allan is paid 5% of her total sales. How much would she earn in a week if her total sales were $2800?

   She would earn _____.

4. Forty percent of the class finished their assignment before lunch. There are 25 pupils in the class. How many pupils finished before lunch?

   _____ pupils finished before lunch.

5. The excise tax on a certain item is 10% of the sales price. What would be the amount of excise tax on an item which sells for $60?

   The excise tax would be _____.

6. It is estimated that a new truck will be worth 75% of its original cost after one year. How much would a 1-year old truck be worth that originally sold for $5600?

   The truck would be worth _____.

7. Fifty percent of the people questioned in a sales survey indicated a preference for Brand X. There were 7,520 people questioned. How many of the people questioned preferred Brand X?

   _____ people preferred Brand X.

Perfect score: 7   My score: _____

Lesson 5 Percent of a Number

Study how decimals are used to find a percent of a number.

```
34% of 62.3  ----→  62.3
    └----------→   ×.34
                   ────
                   2492
                  18690
                  ─────
                  21.182
```

34% of 62.3 = ___21.182___

Complete the following.

                 a   *b*

1. 28% of 62.5 = _____   7.5% of 34 = _____

2. 73% of 95 = _____   5.5% of 9.6 = _____

3. 9.5% of 780 = _____   2% of 73.6 = _____

4. 5% of 8.5 = _____   39% of 420 = _____

5. 6.25% of 700 = _____   125% of 62.5 = _____

6. 85% of 672 = _____   7% of 86.4 = _____

7. 9% of 960 = _____   10% of 95.6 = _____

8. 140% of 280 = _____   8.5% of 785 = _____

9. 25% of 386 = _____   18% of 70 = _____

10. 67% of 18.5 = _____   7.75% of 62.4 = _____

11. 107% of 600 = _____   8% of 420 = _____

12. 83% of 840 = _____   106% of 780 = _____

Perfect score: 24   My score: _____

Problem Solving

Solve each problem.

1. During the sale Mr. Hansen purchased a coat for 60% off the regular price. The coat normally sold for $220. How much money did he save by buying the coat on sale?

   He saved $ _____.

2. Mrs. James purchased a pair of gloves for 50% off the regular price of $12.50. How much did she pay for the gloves?

   She paid $ _____.

3. During the sale, ladies' coats are selling for 75% of the original price. The original price is $98. What is the sale price of the coats?

   The sale price is _____.

4. A sales tax of 5% is charged on all purchases. What is the sales tax on a purchase of $78?

   The sales tax is _____.

5. Charge-account customers must pay a finance charge of 1.5% of their unpaid balance. What is the finance charge to a customer who has an unpaid balance of $82?

   The finance charge is _____.

Perfect score: 5   My score: _____

NAME _____

CHAPTER 11 TEST

Complete the following. Write each fraction in simplest form.

	fraction	decimal	percent
1.	$\frac{3}{100}$	_____	_____
2.	$\frac{1}{4}$	_____	_____
3.	$\frac{7}{20}$	_____	_____
4.	_____	.06	_____
5.	_____	.39	_____
6.	_____	.125	_____
7.	_____	_____	5%
8.	_____	_____	28%
9.	_____	_____	75%
10.	_____	_____	90%

Complete the following.

11.   25% of   64 = _____

12.   80% of   78 = _____

13.   6.25% of 700 = _____

14.   32.5% of 62.4 = _____

15.   8.5% of 96.8 = _____

Perfect score: 25   My score: _____

NAME _____

Chapter 12

PRE-TEST—Geometry

Match each figure with its name. You will not use all of the letters.

1. _____

a. line

b. line segment

c. ray

d. right angle

2. _____

e. acute angle

f. obtuse angle

g. right triangle

h. rectangle

3. _____

i. triangle

4. _____

5. _____

6. _____

Perfect score: 6   My score: _____

138

Lesson 1 Lines, Line Segments, and Rays

Line AB (denoted $\overleftrightarrow{AB}$) names the line which passes through points A and B. Notice that $\overleftrightarrow{AB}$ and $\overleftrightarrow{BA}$ name the same line.

Line segment CD (denoted $\overline{CD}$) consists of points C and D and all points on the line between C and D. Notice that $\overline{CD}$ and $\overline{DC}$ name the same line segment.

Ray EF (denoted $\overrightarrow{EF}$) consists of point E and all points on $\overleftrightarrow{EF}$ that are on the same side of E as F. Notice that $\overrightarrow{EF}$ and $\overrightarrow{FE}$ do **not** name the same ray.

Complete the following as shown.

		a	b

1.    line __JW__ or __WJ__ $\overleftrightarrow{JW}$ or $\overleftrightarrow{WJ}$

2.    ray _____   _____

3.    line segment _____ or _____   _____ or _____

4.    line segment _____ or _____   _____ or _____

5.    line _____ or _____   _____ or _____

6. 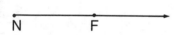   ray _____   _____

7.    ray _____   _____

8.    line segment _____ or _____   _____ or _____

Perfect score: 22   My score: _____

139

NAME _____

Lesson 2 Angles

An **angle** is formed by two rays which have a common endpoint. Angle RTS (denoted ∠RTS) is formed by ray TR and ray TS.

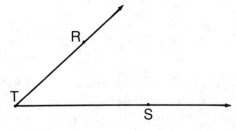

Does ∠STR name the same angle as ∠RTS? _____

You can find the measure of an angle with a protractor.

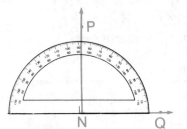   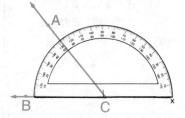   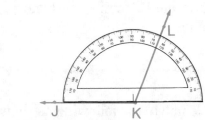

| If the measure of an angle is 90°, the angle is a **right angle**. | If the measure of an angle is less than 90°, the angle is an **acute angle**. | If the measure of an angle is greater than 90°, the angle is an **obtuse angle**. |

Name each angle. Find the measure of each angle. Tell whether the angle is right, acute, or obtuse.

                                                     a   *b*   *c*

1. 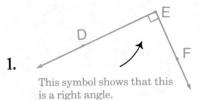 ∠_____ or ∠_____   _____° _____
   This symbol shows that this is a right angle.

2. 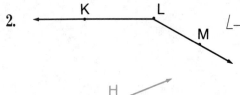 ∠_____ or ∠_____   _____° _____

3.  ∠_____ or ∠_____   _____° _____

4. 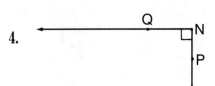 ∠_____ or ∠_____   _____° _____

5. 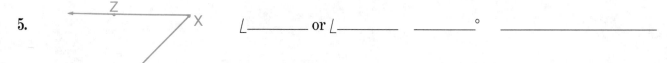 ∠_____ or ∠_____   _____° _____

Perfect score: 20   My score: _____

Lesson 3 Triangles and Quadrilaterals

  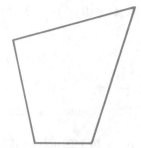 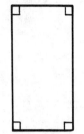 

A **triangle** has 3 sides.

A **right triangle** is a triangle that has 1 right angle.

A **quadrilateral** has 4 sides.

A **rectangle** is a quadrilateral that has 4 right angles.

A **square** is a rectangle that has 4 sides that are all the same length.

Use the figures below to answer each question. You may use some letters more than once. You may not use all of the letters.

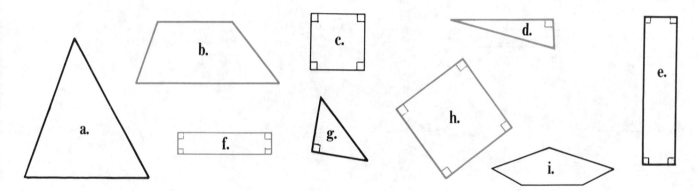

1. Which figures are triangles? _____

2. Which figures are right triangles? _____

3. Which figures are quadrilaterals? _____

4. Which figures are rectangles? _____

5. Which figures are squares? _____

6. Which figures are triangles, but not right triangles? _____

7. Which figures are quadrilaterals, but are not rectangles? _____

Perfect score: 7   My score: _____

NAME _____

CHAPTER 12 TEST

Match each figure with its name. You will not use all of the letters.

1. _____   ———————→

2. _____   ←———•———•———→

3. _____   [rectangle figure]

4. _____   [obtuse angle figure]

5. _____   •———————•

6. _____   [right triangle figure]

7. _____   [acute angle figure]

a. acute angle
b. right triangle
c. right angle
d. rectangle
e. line
f. ray
g. line segment
h. obtuse angle

Perfect score: 7   My score: _____

142

TEST—Chapters 1–6

Complete as indicated.

	a	b	c	d	e
1.	438 +629	82564 7955 +90651	4265 −3051	37128 −18067	97824 −86987
2.	217 ×8	1348 ×6	234 ×75	6954 ×914	8261 ×803
3.	47)564	9)26741	35)8206	20)83275	86)17824
4.	4.6 +7.8	.867 .241 +.378	$27.65 +18.20	$12.65 −9.36	18.27 −9.658

Write each answer in simplest form.

	a	b	c	d	e
5.	$\frac{1}{4} \times \frac{5}{6}$	$\frac{3}{5} \times \frac{10}{21}$	$8 \times \frac{2}{3}$	$2 \times 3\frac{1}{2}$	$2\frac{6}{7} \times 1\frac{2}{5}$
6.	$\frac{4}{5}$ $+\frac{1}{5}$	$\frac{1}{2}$ $+\frac{2}{3}$	$\frac{7}{8}$ $+\frac{5}{6}$	$3\frac{1}{3}$ $+2\frac{1}{4}$	$4\frac{1}{2}$ $2\frac{2}{3}$ $+5\frac{3}{4}$

Continued on the next page.

TEST—Chapters 1–6 (Continued)

Complete the following so the numerals in each row name the same number.

	fractions or mixed numerals	decimals		
		tenths	hundredths	thousandths
7.			2.50	
8.	$4\frac{1}{10}$			

Write each answer in simplest form.

	a	b	c	d	e
9.	$\frac{6}{7}$ $-\frac{4}{7}$	$\frac{9}{10}$ $-\frac{3}{5}$	8 $-\frac{7}{8}$	$6\frac{3}{4}$ $-4\frac{1}{2}$	$5\frac{1}{3}$ $-3\frac{5}{8}$
10.	$8 \div \frac{1}{3}$	$\frac{2}{5} \div 7$	$\frac{7}{8} \div \frac{3}{4}$	$1\frac{1}{2} \div 3$	$2\frac{1}{4} \div 1\frac{1}{2}$

Solve each problem. Write each answer in simplest form.

11. A truck was carrying $\frac{3}{4}$ ton of sand. Two thirds of the sand was used to make cement. How much sand was used to make cement?

_____ ton of sand was used to make cement.

12. Last month $1\frac{1}{2}$ inches of rain fell. This month $3\frac{1}{4}$ inches of rain fell. How much more rain fell this month than last month?

_____ inches more rain fell this month.

13. Ranita bought a purse for $27.65, a blouse for $21.89, and a pair of shoes for $58.80. The sales tax was $5.42. How much did she spend in all?

She spent $ _____ in all.

14. Each sheet of metal is .024 centimeter thick. There are 136 sheets of metal in a stack. How high is the stack?

The stack is _____ centimeters tall.

Perfect score: 50   My score: _____

FINAL TEST—Chapters 1–12

Complete as indicated.

	a	b	c	d	e
1.	5627 +1795	452 1653 +7659	71256 15234 +6954	295 −176	6243 −812
2.	9006 −7328	87240 −15967	178 ×9	267 ×38	8765 ×27
3.	1074 ×465	6)295	8)13765	12)4380	78)86534
4.	6.7 +3.24	8.917 3.1 +16.27	.507 +.4982	25.68 3.5 +7.92	.87 −.6
5.	2.07 −1.695	15.346 −9.29	.215 ×.6	243 ×.65	7.15 ×1.3
6.	1.8 ×.005	3)12.45	.8)3.4	1.3)3.12	.009).81

Write each answer in simplest form.

	a	b	c	d
7.	$\frac{1}{3} \times \frac{3}{5}$	$7 \times \frac{3}{5}$	$5\frac{1}{2} \times 4$	$3\frac{1}{5} \times 6\frac{1}{4}$

Continued on the next page.

Final Test (Continued)

Write each answer in simplest form.

	a	b	c	d
8.	$\frac{7}{8} + \frac{3}{8}$	$\frac{5}{6} + \frac{2}{3}$	$3\frac{1}{2} + 4\frac{1}{4}$	$5\frac{3}{4} + 6\frac{5}{6}$
9.	$\frac{9}{10} - \frac{3}{10}$	$\frac{5}{8} - \frac{1}{4}$	$7\frac{1}{2} - 3\frac{1}{3}$	$6\frac{1}{8} - 4\frac{5}{6}$
10.	$9 \div \frac{1}{4}$	$\frac{7}{8} \div 6$	$\frac{3}{7} \div \frac{9}{10}$	$3\frac{3}{4} \div 1\frac{2}{3}$

Find the area of each figure.

a

11.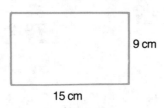

_____ square centimeters

b

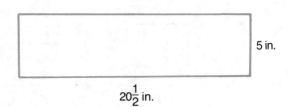

_____ square inches

Find the volume of each figure.

a

12.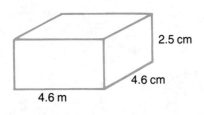

_____ cubic meters

b

_____ cubic inches

Continued on the next page.

Final Test (Continued)

Complete the following. Write each fraction in simplest form.

	percent	fraction	decimal
13.	10%		
14.		$\frac{3}{4}$	
15.			.5

                         a   b

16.   8% of 75 = _____   136% of 10.5 = _____

17.   3.2% of 800 = _____   95% of 400 = _____

Match each colored figure with its name. You will not use all the letters.

18. _____   **a.** acute angle

b. ray

19. _____   **c.** line

d. line segment

20. _____   **e.** obtuse angle

f. right triangle

21. _____

g. right angle

22. _____

Continued on the next page.

Final Test (Continued)

Complete the following.

	a	b
23.	2.3 m = _____ cm	235 cm = _____ mm
24.	2 kl = _____ liters	678 g = _____ kg
25.	3520 yd = _____ mi	3 yd 4 in. = _____ in.
26.	6 cups = _____ pt	4 gal = _____ qt
27.	5 lb = _____ oz	5 hr 15 min = _____ min

Solve. Write each answer in simplest form.

28. Will gave the clerk a twenty-dollar bill to pay for items that totaled $15.34. How much change should he get?

He should get $ _____ in change.

29. A machine can make 382 parts each hour. The machine runs 80 hours each week. How many parts can that machine make in a week?

The machine can make _____ parts in a week.

30. Joanne can make one widget in $1\frac{1}{2}$ hours. It takes Ben $1\frac{1}{4}$ times as long to make one widget. How long does it take Ben to make one widget?

It takes _____ hours for Ben to make one widget.

31. Last month the Build-It Company made 32,626 widgets. The same number of widgets were made each day. The Build-It Company was operating 22 days last month. How many widgets were made each day?

There were _____ widgets made each day.

32. Merilynne worked the following number of hours last week: $7\frac{1}{2}$, $4\frac{3}{4}$, $8\frac{3}{4}$, $7\frac{1}{2}$, $9\frac{1}{4}$. How many hours in all did she work last week?

She worked _____ hours last week.

Perfect score: 80   My score: _____

Answers
Math - Grade 6

(Answers for Pre-Tests and Tests are given on pages 157–159.)

Page 3

	a	b	c	d	e	f
1.	39	27	53	45	59	87
2.	59	78	78	78	69	78
3.	81	63	77	146	103	148
4.	124	151	131	111	102	152
5.	59	80	115	127	68	124
6.	186	134	109	178	180	216

Page 4

	a	b	c	d	e	f
1.	54	94	60	69	29	38
2.	43	43	54	25	10	21
3.	25	48	34	19	7	17
4.	127	156	268	107	226	59
5.	70	191	286	93	390	485
6.	87	283	376	79	437	549

Page 5

	a	b	c	d	e
1.	588	846	858	1267	1422
2.	6408	8343	9372	8733	11356
3.	78553	81378	97121	91967	91170
4.	1262	11411	12241	91272	73549
5.	60321	80121	9023	46838	53757
6.	1684	15287	13564	104566	103743
7.	42869	48277	76063	85022	79379

Page 6

1. 528 ; 746 ; 1274
2. 5281 ; 7390 ; 12671
3. 42165 ; 34895 ; 77060
4. 15342
5. 94400
6. 86889

Page 7

	a	b	c	d	e
1.	412	217	351	179	404
2.	3902	1929	4895	1889	849
3.	3031	1434	2088	2314	5048
4.	48913	39516	48917	53750	24321
5.	20937	23476	38708	34809	12299
6.	26035	23761	65995	60693	441

Page 8

1. 500 ; 385 ; 115
2. 1516 ; 842 ; 674
3. 1464
4. 48459
5. 36332
6. 13687

Page 9

	a	b	c	d	e	f
1.	38	53	58	73	139	135
2.	55	51	9	66	58	165
3.	887	965	877	661	1112	1622
4.	281	508	281	1788	1788	998
5.	8699	6840	10087	11324	11040	
6.	5114	4079	5670	3883	16809	
7.	73545	87550	75520	90417	95732	
8.	42101	48075	10244	29289	7718	
9.	99	1493	11796	88088	84788	

Page 10

1. add ; 1687
2. subtract ; 427
3. add ; 3043
4. subtract ; 3695
5. add ; 61357
6. add ; 60114

Page 13

	a	b	c	d	e	f	g	h
1.	0	0	0	0	7	6	1	5
2.	16	4	8	14	12	10	6	18
3.	27	21	15	0	3	18	12	9

Page 13 (continued)

	a	b	c	d	e	f	g	h
4.	16	12	20	32	28	0	36	4
5.	40	10	35	25	20	15	5	0
6.	48	12	54	42	36	30	6	18
7.	63	49	42	0	7	35	56	28
8.	0	40	64	72	32	24	48	56
9.	27	81	72	9	18	63	54	36

Page 14

	a	b	c	d	e	f	g	h
1.	2	3	5	4	6	9	8	7
2.	9	6	7	8	4	5	2	1
3.	0	5	3	4	8	6	1	7
4.	5	2	1	3	8	6	9	4
5.	6	9	0	2	5	3	8	1
6.	5	4	7	1	2	6	9	8
7.	0	3	2	8	7	9	5	4
8.	2	0	7	9	6	4	3	5
9.	5	3	4	7	1	9	0	6

Page 15

	a	b	c	d	e
1.	96	92	246	624	570
2.	842	492	723	2505	954
3.	2349	4304	3810	6678	4257
4.	2066	9648	9516	10028	9771
5.	8868	40166	43434	32304	35415

Page 16

1. 24 ; 5 ; 120
2. 77 ; 7 ; 539
3. 365 ; 3 ; 1095
4. 3875
5. 6802
6. 20755

Page 17

	a	b	c	d	e
1.	1197	1536	4725	6480	11592
2.	14337	76820	50328	172044	651636
3.	88382	300048	90272	537536	
4.	1243380	2411010	1889280	3449412	

Page 18

1. 28 ; 35 ; 980
2. 47 ; 19 ; 893
3. 321 ; 52 ; 16692
4. 104224
5. 105264
6. 525600

Page 19

	a	b	c	d	e
1.	23	19 r1	24	19 r1	135
2.	75	98 r2	1346 r4	526 r5	709

Page 20

1. 92 ; 5 ; 18
2. 258
3. 305
4. 68 ; 1
5. 3258
6. 384 ; 3

Page 21

	a	b	c	d	e
1.	32 r10	27	45 r9	26 r25	92
2.	142 r27	346 r10	356 r49	525 r25	351

Page 22

1. 988 ; 26 ; 38
2. 41 ; 3
3. 309 ; 25
4. 75
5. 225
6. 752 ; 28

Page 23

	a	b	c	d	e
1.	245	1735	3080	12942	29436
2.	3145	2592	32844	10982	117612
3.	20856	222390	76245	526787	577382
4.	17 r12	12 r32	38	122 r4	157
5.	206 r2	63 r3	425	468 r11	1062 r17

Page 24
1. multiply ; 48750
2. divide ; 7
3. multiply ; 43680
4. divide ; 77 ; 15
5. divide ; 123 ; 10
6. multiply ; 524160

Page 27

	a	b	c	d		a	b
1.	$\frac{1}{2}$	$\frac{3}{4}$	$\frac{1}{3}$	$\frac{1}{5}$	6.	$\frac{4}{5}$	$\frac{7}{8}$
2.	$\frac{3}{5}$	$\frac{3}{8}$	$\frac{5}{6}$	$\frac{5}{8}$	7.	$\frac{5}{6}$	$\frac{7}{9}$
	a	b			8.	$\frac{4}{7}$	$\frac{3}{5}$
3.	$\frac{1}{2}$	$\frac{3}{8}$			9.	$\frac{5}{8}$	$\frac{2}{7}$
4.	$\frac{2}{3}$	$\frac{4}{7}$			10.	$\frac{9}{10}$	$\frac{4}{9}$
5.	$\frac{3}{4}$	$\frac{3}{7}$					

Page 28

	a	b	c	d		a	b	c
1.	$\frac{1}{5}$	4	3 ; $\frac{3}{4}$	$9\frac{1}{3}$	4.	$2\frac{1}{2}$	$1\frac{4}{5}$	$3\frac{1}{2}$
2.	$\frac{2}{3}$	5	6 ; $\frac{2}{5}$	$8\frac{7}{8}$	5.	$2\frac{1}{4}$	$1\frac{1}{5}$	$2\frac{2}{3}$
3.	$\frac{1}{8}$	2	3 ; $\frac{1}{3}$	$5\frac{3}{7}$	6.	$4\frac{2}{3}$	$3\frac{1}{3}$	$3\frac{2}{5}$

	a	b	c
7.	less than 1	greater than 1	equal to 1
8.	less than 1	equal to 1	greater than 1
9.	less than 1	greater than 1	greater than 1

Page 29

	a	b	c	d
1.	$\frac{4}{5}$	$\frac{7}{8}$	$\frac{4}{7}$	$\frac{4}{5}$
2.	$\frac{5}{6}$	$\frac{4}{7}$	$\frac{3}{8}$	$\frac{3}{4}$
3.	$\frac{7}{10}$	$\frac{5}{12}$	$\frac{9}{11}$	$\frac{11}{15}$

	a	b	c	d	e	f
4.	$\frac{5}{6}$	$\frac{7}{8}$	$\frac{3}{7}$	$\frac{9}{10}$	$\frac{11}{12}$	$\frac{4}{11}$
5.	$\frac{3}{5}$	$\frac{6}{7}$	$\frac{5}{8}$	$\frac{7}{10}$	$\frac{11}{15}$	$\frac{7}{12}$

Page 30

	a	b	c		a	b	c
1.	$\frac{21}{8}$	$\frac{13}{5}$	$\frac{11}{3}$	3.	$\frac{55}{8}$	$\frac{59}{10}$	$\frac{161}{12}$
2.	$\frac{37}{10}$	$\frac{32}{3}$	$\frac{29}{2}$	4.	$\frac{29}{6}$	$\frac{31}{4}$	$\frac{107}{12}$

Page 31

	a	b	c	d
1.	$\frac{1}{6}$	$\frac{3}{8}$	$\frac{1}{12}$	$\frac{3}{10}$
2.	$\frac{9}{20}$	$\frac{12}{35}$	$\frac{8}{15}$	$\frac{15}{56}$
3.	$\frac{8}{15}$	$\frac{1}{16}$	$\frac{15}{28}$	$\frac{21}{40}$
4.	$\frac{18}{35}$	$\frac{2}{27}$	$\frac{15}{56}$	$\frac{6}{35}$
5.	$\frac{49}{64}$	$\frac{4}{9}$	$\frac{8}{27}$	$\frac{24}{35}$
6.	$\frac{40}{63}$	$\frac{5}{24}$	$\frac{25}{42}$	$\frac{15}{64}$

Page 32

	a	b	c	d
1.		4	8	9
2.		24	30	25
3.	8	80	10	15

Page 33
1. 1,2,3,6 ; 1,2 ; 2
   1,2,5,10
2. 1,5 ; 1 ; 1
   1,2,4,8
3. 1,2,3,4,6,12 ; 1,3 ; 3
   1,3,5,15

Page 33 (continued)
4. 1,2,5,10 ; 1,2,5,10 ; 10
   1,2,4,5,10,20
5. 1,2,7,14 ; 1,2 ; 2
   1,2,4,8,16
6. 1,3,5,15 ; 1 ; 1
   1,7
7. 1,2,3,4,6,8,12,24 ; 1,2,3,6 ; 6
   1,2,3,6,9,18

Page 34

	a	b	c		a	b	c
1.	$\frac{4}{5}$	$\frac{1}{2}$	$\frac{2}{3}$	4.	$6\frac{2}{3}$	$\frac{5}{6}$	$3\frac{3}{4}$
2.	$2\frac{1}{2}$	$3\frac{2}{3}$	$5\frac{4}{5}$	5.	$\frac{4}{5}$	$3\frac{7}{9}$	$\frac{1}{2}$
3.	$\frac{2}{3}$	$5\frac{3}{4}$	$\frac{5}{6}$				

Page 35

	a	b	c	d
1.	$\frac{3}{10}$	$\frac{8}{15}$	$\frac{4}{9}$	$\frac{5}{42}$
2.	$\frac{3}{5}$	$\frac{5}{9}$	$\frac{4}{7}$	$\frac{4}{15}$
3.	$\frac{1}{3}$	$\frac{2}{3}$	$\frac{1}{4}$	$\frac{1}{6}$
4.	$\frac{9}{20}$	$\frac{3}{8}$	$\frac{4}{15}$	$\frac{3}{4}$
5.	$\frac{10}{21}$	$\frac{21}{80}$	$\frac{1}{2}$	$\frac{27}{35}$

Page 36
1. $\frac{3}{8}$ 3. $\frac{2}{5}$ 5. $\frac{1}{4}$ 7. $\frac{1}{6}$
2. $\frac{8}{15}$ 4. $\frac{1}{2}$ 6. $\frac{1}{3}$

Page 37

	a	b	c	d
1.	$3\frac{1}{3}$	$4\frac{4}{5}$	$4\frac{1}{2}$	$5\frac{1}{4}$
2.	$7\frac{1}{2}$	$1\frac{1}{2}$	$4\frac{1}{2}$	8
3.	$7\frac{1}{2}$	$6\frac{2}{3}$	$6\frac{4}{5}$	$18\frac{2}{3}$

Page 38
1. 2   3. $10\frac{1}{3}$ 5. 3   7. 16 ; 8
2. 8   4. $9\frac{3}{8}$ 6. $1\frac{1}{2}$

Page 39

	a	b	c	d		a	b	c	d
1.	$6\frac{8}{15}$	$4\frac{1}{12}$	$4\frac{1}{6}$	$7\frac{1}{9}$	3.	$8\frac{3}{4}$	$6\frac{3}{8}$	$5\frac{1}{4}$	$12\frac{1}{2}$
2.	$5\frac{2}{5}$	$4\frac{1}{4}$	$6\frac{2}{5}$	$3\frac{1}{5}$	4.	$32\frac{15}{16}$	$6\frac{1}{2}$	$1\frac{9}{16}$	$20\frac{5}{6}$

Page 40
1. $3\frac{5}{9}$ 3. $31\frac{7}{8}$ 5. $15\frac{3}{4}$ 7. 26   9. $3\frac{3}{4}$
2. 6   4. $4\frac{1}{12}$ 6. $25\frac{1}{3}$ 8. $5\frac{1}{4}$

Page 43

	a	b	c	d	e
1.	$\frac{3}{5}$	$\frac{6}{7}$	$1\frac{1}{4}$	$1\frac{1}{2}$	$1\frac{3}{4}$
2.	$\frac{1}{6}$	$\frac{1}{2}$	$\frac{3}{7}$	$\frac{5}{9}$	$\frac{1}{2}$
3.	$\frac{9}{10}$	$1\frac{1}{3}$	$\frac{3}{4}$	$\frac{5}{6}$	$1\frac{3}{5}$
4.	$\frac{2}{3}$	$\frac{5}{8}$	$\frac{1}{3}$	$\frac{1}{2}$	$\frac{3}{8}$
5.	$1\frac{1}{4}$	$\frac{1}{3}$	$1\frac{1}{5}$	$\frac{2}{5}$	$\frac{6}{7}$

Page 44
1. $\frac{1}{2}$ 3. $\frac{3}{4}$ 5. $\frac{1}{6}$ 7. 1
2. $\frac{1}{2}$ 4. $\frac{5}{8}$ 6. $1\frac{1}{2}$ 8. $\frac{2}{3}$

Answers Grade 6

Page 45

	a	b	c	d		a	b	c	d
1.	$1\frac{4}{15}$	$1\frac{1}{30}$	$\frac{5}{6}$	$\frac{19}{30}$	3.	$1\frac{5}{24}$	$\frac{13}{24}$	$1\frac{3}{20}$	$\frac{1}{6}$
2.	$\frac{5}{12}$	$\frac{13}{30}$	$\frac{5}{24}$	$\frac{5}{12}$	4.	$1\frac{1}{12}$	$\frac{4}{15}$	$1\frac{3}{10}$	$\frac{1}{12}$

Page 46

	a	b	c	d
1.	$\frac{9}{10}$	$\frac{11}{15}$	$1\frac{1}{4}$	$1\frac{1}{8}$
2.	$\frac{5}{12}$	$\frac{1}{10}$	$\frac{1}{2}$	$1\frac{1}{5}$
3.	$1\frac{2}{5}$	$\frac{1}{12}$	$1\frac{1}{3}$	$1\frac{1}{2}$
4.	$1\frac{1}{2}$	$\frac{2}{3}$	$1\frac{2}{15}$	$2\frac{2}{5}$

Page 47

	a	b	c	d
1.	$6\frac{1}{20}$	$3\frac{11}{12}$	$7\frac{1}{8}$	$4\frac{3}{4}$
2.	$10\frac{5}{8}$	$11\frac{7}{10}$	$3\frac{13}{30}$	$4\frac{11}{15}$
3.	$8\frac{7}{12}$	$5\frac{19}{20}$	$7\frac{3}{4}$	7
4.	$4\frac{23}{30}$	$6\frac{5}{24}$	$6\frac{5}{8}$	$8\frac{17}{30}$

Page 48

	a	b	c	d
1.	$6\frac{1}{4}$	$3\frac{1}{2}$	$4\frac{1}{3}$	$7\frac{7}{8}$
2.	$2\frac{3}{10}$	$2\frac{2}{9}$	$3\frac{1}{3}$	$2\frac{1}{2}$
3.	$4\frac{2}{5}$	$1\frac{5}{8}$	$1\frac{1}{6}$	$7\frac{7}{10}$
4.	$3\frac{5}{12}$	$7\frac{3}{8}$	$7\frac{1}{2}$	$5\frac{1}{10}$

Page 49

	a	b	c	d
1.	$\frac{3}{4}$	$\frac{11}{12}$	$3\frac{11}{12}$	$1\frac{9}{10}$
2.	$6\frac{1}{2}$	$\frac{1}{2}$	$1\frac{5}{6}$	$1\frac{19}{24}$
3.	$3\frac{4}{5}$	$2\frac{1}{2}$	$6\frac{3}{4}$	$\frac{4}{5}$
4.	$1\frac{19}{20}$	$5\frac{5}{8}$	$6\frac{5}{6}$	$6\frac{5}{12}$

Page 50

1. $\frac{5}{12}$ 3. $1\frac{1}{4}$ 5. $1\frac{1}{10}$ 7. $1\frac{7}{40}$
2. $\frac{9}{20}$ 4. $\frac{5}{8}$ 6. $\frac{1}{10}$ 8. $\frac{23}{40}$

Page 51

	a	b	c	d
1.	$1\frac{1}{2}$	$\frac{3}{8}$	1	$\frac{2}{3}$
2.	$1\frac{3}{8}$	$\frac{2}{15}$	$1\frac{5}{18}$	$\frac{1}{3}$
3.	$8\frac{4}{9}$	$\frac{7}{10}$	$5\frac{1}{8}$	$8\frac{1}{4}$
4.	$4\frac{7}{15}$	$6\frac{17}{24}$	$6\frac{7}{12}$	$\frac{7}{10}$
5.	$10\frac{5}{6}$	$14\frac{3}{4}$	$10\frac{29}{120}$	$8\frac{1}{3}$

Page 52

1. Jennifer; $\frac{1}{12}$ 4. $\frac{3}{4}$
2. $\frac{1}{2}$ 5. $4\frac{11}{12}$
3. $2\frac{1}{6}$ 6. $1\frac{1}{3}$

Page 55

	a	b	c	d	e	f
1.	$\frac{5}{3}$	$\frac{8}{7}$	$\frac{5}{4}$	$\frac{7}{5}$	$\frac{9}{4}$	$\frac{7}{6}$
2.	$\frac{3}{5}$	$\frac{7}{8}$	$\frac{4}{5}$	$\frac{5}{7}$	$\frac{4}{9}$	$\frac{6}{7}$
3.	8	3	4	9	16	14

Page 55 (continued)

	a	b	c	d	e	f
4.	$\frac{1}{8}$	$\frac{1}{3}$	$\frac{1}{4}$	$\frac{1}{9}$	$\frac{1}{16}$	$\frac{1}{14}$
5.	$\frac{1}{8}$	$\frac{1}{3}$	$\frac{1}{4}$	$\frac{1}{9}$	$\frac{1}{16}$	$\frac{1}{14}$
6.	$\frac{5}{8}$	$\frac{1}{6}$	$\frac{3}{2}$	$\frac{6}{11}$	$\frac{4}{7}$	$\frac{1}{12}$
7.	$\frac{1}{15}$	$\frac{9}{10}$	$\frac{11}{12}$	$\frac{1}{17}$	$\frac{9}{8}$	$\frac{2}{17}$
8.	$\frac{8}{15}$	$\frac{12}{5}$	$\frac{1}{11}$	$\frac{11}{7}$	11	$\frac{3}{17}$
9.	$\frac{1}{10}$	$\frac{1}{13}$	17	$\frac{11}{5}$	$\frac{7}{9}$	$1\frac{5}{2}$
10.	$\frac{8}{5}$	6	$\frac{1}{7}$	$\frac{7}{12}$	$\frac{1}{2}$	$\frac{5}{2}$

Page 56

	a	b	c	d
1.	30	16	28	30
2.	49	$37\frac{1}{2}$	$42\frac{2}{3}$	$32\frac{2}{5}$
3.	54	16	34	16

Page 57

	a	b	c	d
1.	$\frac{1}{12}$	$\frac{1}{8}$	$\frac{1}{15}$	$\frac{1}{12}$
2.	$\frac{3}{20}$	$\frac{5}{16}$	$\frac{3}{16}$	$\frac{5}{18}$
3.	$\frac{1}{8}$	$\frac{1}{9}$	$\frac{1}{5}$	$\frac{1}{12}$

Page 58

1. $\frac{1}{6}$ 3. $\frac{1}{4}$ 5. $\frac{7}{32}$
2. $\frac{1}{8}$ 4. $\frac{1}{6}$ 6. $\frac{1}{6}$

Page 59

	a	b	c	d
1.	$\frac{2}{5}$	$\frac{2}{3}$	$\frac{1}{2}$	$\frac{2}{3}$
2.	$1\frac{1}{5}$	$\frac{6}{7}$	8	$1\frac{1}{4}$
3.	2	$\frac{1}{2}$	$2\frac{2}{9}$	$1\frac{1}{4}$

Page 60

1. 3   3. 6   5. 4   7. 5   9. 4
2. 3   4. $4\frac{1}{2}$ 6. 2   8. 2

Page 61

	a	b	c	d		a	b	c	d
1.	12	$\frac{4}{35}$	$\frac{1}{2}$	$\frac{9}{10}$	4.	$\frac{1}{5}$	2	$1\frac{1}{4}$	$7\frac{1}{2}$
2.	$1\frac{1}{4}$	24	$\frac{1}{6}$	$\frac{3}{7}$	5.	1	$\frac{27}{28}$	21	$\frac{3}{14}$
3.	$2\frac{1}{2}$	$1\frac{1}{8}$	15	$\frac{2}{27}$

Page 62

1. 27   3. 2   5. $\frac{1}{8}$ 7. $\frac{1}{16}$
2. $\frac{1}{12}$ 4. 12   6. $\frac{1}{2}$ 8. $\frac{3}{20}$

Page 63

	a	b	c	d
1.	$\frac{5}{6}$	$\frac{7}{15}$	3	$4\frac{1}{2}$
2.	$\frac{18}{35}$	$\frac{9}{20}$	$3\frac{3}{4}$	$1\frac{1}{2}$
3.	$6\frac{3}{10}$	$\frac{1}{9}$	$\frac{9}{25}$	$\frac{8}{15}$

Page 64

1. 4   3. 3   5. 9   7. 6
2. 10   4. 6   6. 4

Page 67

	a	b	c	d
1.	.6	.2	.8	.5
2.	4.7	5.9	18.2	423.6
3.	$\frac{7}{10}$	$\frac{3}{10}$	$\frac{1}{10}$	$\frac{9}{10}$
4.	$4\frac{9}{10}$	$12\frac{7}{10}$	$15\frac{1}{10}$	$217\frac{3}{10}$
5.	.8	3.7		
6.	.4	25.8		
7.	.5	100.6		

8. nine tenths
9. three and seven tenths
10. twenty-one and two tenths

Page 68

	a	b	c
1.	.08	.16	.05
2.	1.36	8.06	9.12
3.	12.45	43.67	26.04
4.	142.08	436.42	389.89
5.	$\frac{17}{100}$	$\frac{3}{100}$	$\frac{41}{100}$
6.	$5\frac{19}{100}$	$6\frac{47}{100}$	$5\frac{1}{100}$
7.	$21\frac{7}{100}$	$23\frac{99}{100}$	$44\frac{89}{100}$
8.	$142\frac{33}{100}$	$483\frac{3}{100}$	$185\frac{63}{100}$

	a	b
9.	.08	6.23
10.	.95	14.60
11.	.48	4.44

Page 69

1.	.008	.017	.0054
2.	.0125	.430	.0306
3.	4.004	3.0041	6.183
4.	35.0078	42.019	196.006
5.	$\frac{9}{1000}$	$\frac{19}{10000}$	$\frac{3}{10000}$
6.	$\frac{123}{1000}$	$\frac{441}{10000}$	$\frac{219}{1000}$
7.	$4\frac{11}{1000}$	$2\frac{1011}{10000}$	$6\frac{14}{1000}$
8.	$36\frac{37}{1000}$	$3\frac{433}{1000}$	$100\frac{1}{10000}$

	a	b
9.	.053	10.009
10.	.0011	12.018
11.	.065	12.001

Page 70

	a	b	c
1.	.6	.60	.600
2.	3.5	.28	2.190

	a	b	c
3.	2.8	.35	.056
4.	2.2	.38	.352

Page 71

	a	b	c	d
1.	$\frac{3}{10}$	$\frac{1}{10}$	$\frac{2}{5}$	$\frac{1}{2}$
2.	$2\frac{7}{10}$	$3\frac{3}{10}$	$7\frac{1}{5}$	$5\frac{4}{5}$
3.	$\frac{17}{100}$	$\frac{3}{100}$	$\frac{3}{20}$	$\frac{4}{5}$
4.	$5\frac{7}{100}$	$8\frac{43}{100}$	$4\frac{1}{20}$	$2\frac{11}{25}$
5.	$\frac{3}{1000}$	$\frac{17}{1000}$	$\frac{1}{8}$	$\frac{9}{200}$
6.	$3\frac{121}{1000}$	$2\frac{987}{1000}$	$4\frac{1}{4}$	$3\frac{1}{125}$
7.	$4\frac{7}{20}$	$\frac{7}{10}$	$6\frac{1}{5}$	$1\frac{7}{1000}$
8.	$2\frac{3}{5}$	$3\frac{6}{25}$	$\frac{1}{4}$	$3\frac{1}{2}$

Page 71 (continued)

9.	$5\frac{1}{8}$	$\frac{9}{10}$	$2\frac{2}{5}$	$\frac{1}{25}$
10.	$\frac{1}{100}$	$\frac{51}{1000}$	$\frac{4}{5}$	$2\frac{19}{100}$

Page 72

	a	b	c	d
1.	.2	.35	.445	
2.	7.5	4.58	3.360	
3.	$\frac{9}{10}$	$3\frac{3}{5}$	$\frac{7}{20}$	$17\frac{3}{4}$
4.	$\frac{1}{40}$	$8\frac{89}{200}$	$24\frac{61}{100}$	$8\frac{1}{20}$
5.	$\frac{3}{5}$	.6	.60	
6.	$2\frac{7}{10}$	2.7		2.700
7.	$5\frac{2}{5}$		5.40	5.400
8.		3.5	3.50	3.500
9.	$17\frac{9}{10}$		17.90	17.900
10.	$80\frac{4}{5}$	80.8		80.800

Page 73

	a	b	c	d	e
1.	.9	1.7	12.6	32.1	52.4
2.	.77	1.24	7.94	96.81	62.21
3.	.245	1.332	4.339	5.813	43.223
4.	1.8	7.5	9.9	46.8	8.4
5.	1.00	$1.04	10.54	$41.89	$37.19
6.	.721	.696	10.620	6.329	38.114

Page 74

1. 2
2. 1.5
3. 8
4. .63
5. .218
6. 3.9
7. 7.1
8. 16

Page 75

	a	b	c	d	e
1.	1.32	1.23	1.001	1.143	.845
2.	6.058	13.14	4.116	8.333	34.43
3.	1.238	.96	2.48	1.081	.915
4.	7.122	16.837	10.944	11.461	10.104

	a	b
5.	1.71	1.166
6.	.694	.853
7.	1.201	8.024
8.	2.71	26.556

Page 76

1. 1.25
2. 1.50 or 1.5
3. 1.225
4. 1.175
5. $66.55
6. $26.55
7. $43.50
8. 5.495

Page 77

	a	b	c	d	e
1.	.4	.7	.4	.8	.3
2.	.11	.33	.05	.48	$.15
3.	.111	.289	.208	.439	.379
4.	1.4	4.6	4.9	4.9	18.3
5.	3.13	$3.82	3.69	$12.65	$3.77
6.	2.212	2.209	2.812	23.802	11.196
7.	10.4	2.48	7.13	1.955	12.388

Page 78

1. .2
2. .5
3. .23
4. .24
5. 2.775
6. 2.40
7. 1.8

Answers Grade 6

Page 79

	a	b	c	d	e
1.	.52	2.16	2.68	3.216	.646
2.	.113	.293	4.861	2.748	.982
3.	.45	.24	5.18	4.65	7.58
4.	.591	.125	2.044	2.868	10.686
5.	.436	.085	4.408	3.788	10.536
6.	32.085	38.925	39.036	23.85	1.076
7.	37.52	318.79	1.026	78.667	89.397

Page 80

1. 1.2
2. Ms. Williams ; Mr. Karns ; 6.3
3. 36.6
4. 3.5

Page 83

	a	b	c	d	e
1.	6	.6	.06	.006	.6
2.	48	4.8	.48	.048	.48
3.	15	1.5	.15	.015	.015
4.	12	.12	.012	.012	.0012
5.	42	.42	.042	.042	.0042
6.	72	.72	.072	.072	.0072

Page 84

	a	b	c	d
1.	44.8	4.48	.448	.0448
2.	129.6	12.96	1.296	.1296
3.	8.84	.884	.0884	88.4
4.	1.924	192.4	19.24	.1924
5.	7.5	.075	.75	.0075
6.	.48	.48	.048	.0048
7.	2.19	.0219	2.19	.219

Page 85

	a	b	c	d	e
1.	3.5	.6	7.2	2.1	.8
2.	.48	.06	.21	.08	.42
3.	.32	.06	.56	.54	.15
4.	.045	.035	.064	.006	.015
5.	.0024	.0024	.0009	.0056	.0006
6.	.063	.048	.024	.032	.035
7.	.0045	.0081	.0012	.0009	.0025

Page 86

	a	b	c	d	e
1.	56.42	564.2	5642	5642	5.642
2.	1.064	10.64	106.4	.106	1064
3.	2.3	23	230	.23	.23
4.	.08	.8	8	8	80
5.	15	150	1500	1500	1.5

Page 87

	a	b	c	d	e
1.	3.44	.216	.0532	.0162	.736
2.	.0411	38.56	362.8	4.571	2.496
3.	16.05	20.24	.2836	.0842	.4527
4.	1.134	.928	.075	.2226	57.76
5.	.6545	3.528	85.33	.4464	3.9075
6.	75.888	9.2796	73.536	.8164	.5328

Page 88

1. 5.4
2. .168
3. 3
4. 21.05
5. .144
6. .192
7. 3.7996
8. 478.8
9. 411.6

Page 89

	a	b	c	d
1.	.24	.72	.0049	.040
2.	1.215	.1216	.1118	18.75
3.	7.897	1.9992	16.324	4.4469
4.	9.5742	114.552	254.904	34.4258
5.	38.5014	175.417	19.4688	45.0294

Page 90

1. 40.8
2. .336
3. 1.608
4. 206.55
5. 245.025
6. 3270
7. 7.84
8. 4.032

Page 93

	a	b	c	d	e
1.	73	7.3	.73	.073	.0073
2.	1.87	.0027	.033	1.16	4.2
3.	3.5	.042	.0027	6.1	.56

Page 94

1. .2
2. 2.3
3. .003
4. .06
5. .001
6. 18.1
7. .012
8. .0084

Page 95

	a	b	c	d
1.	180	270	510	370
2.	1800	2400	1700	3700
3.	3000	3000	4000	31000

Page 96

1. 180
2. 130
3. 13000
4. 650
5. 240
6. 1250
7. 120
8. 200
9. 320

Page 97

	a	b	c	d
1.	18	2.7	.49	92
2.	14	1.7	2.8	36
3.	4	.6	18	3.7
4.	24	70	1.5	12

Page 98

	a	b	c	d
1.	60	480	230	30
2.	800	50	270	500
3.	370	1700	340	1200
4.	65	70	200	140
5.	250	8100	830	9600

Page 99

	a	b	c		a	b	c
1.	.48	2.9	.017	4.	140	1800	280
2.	150	4000	200	5.	120	28000	1.6
3.	1.4	5300	7.9				

Page 100

1. 18
2. 1 ; .5
3. 1700
4. .65
5. 840
6. 10
7. 100

Page 101

	a	b	c	d
1.	5	60	500	500
2.	3.3	1.8	4.6	.25
3.	.58	2.7	4.2	.032
4.	3000	4.2	.073	1.2

Page 102

1. 8
2. 3 ; .7
3. 80
4. 8
5. 4
6. 5
7. 100
8. .01

Page 103

	a	b	c
1.	8.2	.12	500
2.	3.3	120	76
3.	.47	2.8	9.2

Page 104
1. .059
2. .12
3. .14
4. .35
5. .04
6. .10
7. .16

Page 107
1–3. Have your teacher check your work.
4. 7
5. 5
6. 75
7. 42
8–10. Have your teacher check your work.

Page 108

	a	b	c
1.	45	4.5	.045
2.	16	1.6	.016
3.	29	2.9	.029
4.	51	5.1	.051
5.	63	6.3	.063

	a	b	c
6.	5.4	8000	.234
7.	160	9	5.8
8.	.612	4000	1.3
9.	20	750	.0345
10.	7.07	.5	46.5

Page 109

	a	b
1.	4 ; 2 ; 8	1.5 ; 1.5 ; 2.25
2.	30 ; 15 ; 450	25 ; 10 ; 250
3.	17.4	
4.	20.25	
5.	518.4	

Page 110
1. 11250
2. 27
3. 55.25
4. 8400
5. 1820
6. 119
7. 800000

Page 111

	a	b	c
1.	54	64	162
2.	18.75	894.6	.18
3.	210		
4.	124.2		
5.	614.125		
6.	232.128		

Page 112
1–2. Have your teacher check your work.
3. 2.4
4. 588
5. 262.5

Page 113
1. 5000
2. 24
3. 25
4. 8
5. 96000

Page 114

	a	b
1.	7000	500
2.	.005	4.5
3.	7500	2540
4.	.6	.0075
5.	3400	.3
6.	3000	.3
7.	.0006	24000
8.	.047	750
9.	20000 ; 20	

Page 115
1. 50
2. 1000
3. 200
4. 1000
5. 210

Page 116

	a	b
1.	8000	7500
2.	4.5	.038
3.	6000	.64
4.	50	4500
5.	7	7
6.	5500	210
7.	400	.345
8.	.607	8900
9.	52000	.975
10.	50000	

Page 119

	a	b		
1.	72	5	7.	64
2.	27	4	8.	113
3.	180	4	9.	17
4.	15840	5280	10.	114
5.	15	$1\frac{1}{2}$	11.	6030
6.	10560	8800		

Page 120
1. 7
2. 112
3. 375
4. 51
5. 69
6. 46
7. 32
8. 59
9. 64

Page 121

	a	b	c				
1.	54	44	49	4.	$17\frac{1}{2}$	6.	$8\frac{3}{4}$
2.	$3\frac{3}{4}$	$24\frac{3}{4}$	$37\frac{1}{2}$	5.	$13\frac{1}{2}$	7.	$3\frac{3}{4}$
3.	36						

Page 122
1. 64
2. 140
3. 231
4. 216
5. 9600
6. 100
7. $229\frac{1}{2}$

Page 123

	a	b	c		
1.	48	91	$65\frac{1}{4}$	6.	$115\frac{1}{2}$
2.	$82\frac{1}{2}$	$28\frac{1}{2}$	$56\frac{1}{4}$	7.	$35\frac{3}{4}$
3.	105			8.	$146\frac{1}{4}$
4.	$202\frac{1}{2}$			9.	$55\frac{1}{8}$
5.	$175\frac{1}{2}$			10.	$404\frac{11}{16}$

Page 124
1. 4200
2. 450
3. 3072
4. $37\frac{1}{2}$
5. 25200
6. $87\frac{3}{4}$
7. 2592
8. 48

Page 125

	a	b		
1.	6	4	5.	5
2.	10	5	6.	23
3.	16	$2\frac{3}{4}$	7.	5
4.	6	$7\frac{1}{2}$	8.	19
			9.	15 ; 30 ; 60

Page 126

	a	b		
1.	1152	5	7.	6500
2.	8000	3	8.	165
3.	180	2	9.	108
4.	300	6	10.	260
5.	120	6	11.	56
6.	60	86		

Answers Grade 6

Page 129

1.	$\frac{1}{100}$	.01	8.	$\frac{49}{100}$	.49
2.	$\frac{7}{100}$	.07	9.	$\frac{61}{100}$	.61
3.	$\frac{29}{100}$	.29	10.	$\frac{9}{100}$	.09
4.	$\frac{47}{100}$	.47	11.	$\frac{37}{100}$	.37
5.	$\frac{53}{100}$	.53	12.	$\frac{77}{100}$	.77
6.	$\frac{21}{100}$	.21	13.	$\frac{91}{100}$	.91
7.	$\frac{83}{100}$	.83	14.	$\frac{33}{100}$	.33

Page 130

	a	b	c		a	b	c
1.	$\frac{1}{4}$	$\frac{9}{20}$	$1\frac{3}{5}$	6.	20%	75%	5%
2.	$\frac{13}{20}$	$1\frac{1}{5}$	$\frac{6}{25}$	7.	214%	60%	120%
3.	$\frac{39}{50}$	$\frac{11}{20}$	$2\frac{3}{5}$	8.	90%	28%	225%
4.	$\frac{7}{10}$	$1\frac{11}{25}$	$\frac{43}{50}$	9.	160%	30%	16%
5.	$\frac{19}{20}$	$\frac{2}{5}$	$1\frac{4}{5}$	10.	35%	62%	140%

Page 131

	a	b	c
1.	.135	.37	.0625
2.	.065	.0475	.0275
3.	.07	.625	.085
4.	.325	.0875	.095
5.	.0825	.175	.0375
6.	.0075	.0725	.0175
7.	60%	52%	32.5%
8.	24.75%	80%	65%
9.	14.5%	16.75%	50%
10.	6%	.7%	6.25%
11.	7.5%	.75%	.5%
12.	90%	19%	38.5%

Page 132

1. 75%
2. $\frac{13}{20}$
3. .250
4. 80%
5. $\frac{9}{20}$
6. .615
7. 98.7%
8. 70%

Page 133

	a	b		a	b
1.	$4\frac{1}{2}$	$68\frac{29}{50}$	6.	$2\frac{71}{80}$	128
2.	16	$9\frac{9}{20}$	7.	$99\frac{3}{5}$	$1\frac{31}{50}$
3.	$13\frac{3}{5}$	$18\frac{1}{5}$	8.	$5\frac{11}{50}$	336
4.	$2\frac{2}{5}$	$7\frac{43}{50}$	9.	$5\frac{22}{25}$	225
5.	$36\frac{2}{5}$	$10\frac{1}{5}$	10.	725	$3\frac{21}{25}$

Page 134

1. 33
2. 150
3. $140
4. 10
5. $6
6. $4200
7. 3760

Page 135

	a	b		a	b
1.	17.5	2.55	7.	86.4	9.56
2.	69.35	.528	8.	392	66.725
3.	74.1	1.472	9.	96.5	12.6
4.	.425	163.8	10.	12.395	4.836
5.	43.75	78.125	11.	642	33.6
6.	571.2	6.048	12.	697.2	826.8

Page 136

1. $88
2. $6.25
3. $73.50
4. $3.90
5. $1.23

Page 139

	a	b		a	b
1.	JW or WJ	$\overleftrightarrow{JW}$ or $\overleftrightarrow{WJ}$	5.	AD or DA	$\overleftrightarrow{AD}$ or $\overleftrightarrow{DA}$
2.	BC	$\overrightarrow{BC}$	6.	NF	$\overrightarrow{NF}$
3.	GS or SG	$\overline{GS}$ or $\overline{SG}$	7.	PM	$\overrightarrow{PM}$
4.	ER or RE	$\overline{ER}$ or $\overline{RE}$	8.	KH or HK	$\overline{KH}$ or $\overline{HK}$

Page 140

	a	b	c
1.	DEF or FED	90	right
2.	KLM or MLK	120	obtuse
3.	HGJ or JGH	20	acute
4.	QNP or PNQ	90	right
5.	ZXY or YXZ	45	acute

Page 141

1. a. ; d. ; g.
2. d. ; g.
3. b. ; c. ; e. ; f. ; h.
4. c. ; e. ; f. ; h.
5. c. ; h.
6. a.
7. b.

Photo Credits

Glencoe staff photo, 104; Cameron Mitchell, 52, 80, 136; Larry Molmud, 2

Answers for Pre-Tests and Tests for Grade 6

Page vii

	a	b	c	d	e	f	g	h
1.	10	5	8	10	11	6	3	8
2.	9	16	2	15	7	13	6	2
3.	9	0	7	6	10	5	15	6
4.	9	4	14	1	4	14	12	11
5.	8	16	7	11	8	17	3	13
6.	12	7	12	13	13	7	15	12
7.	8	11	14	8	16	11	12	17
8.	9	14	4	12	9	11	12	5
9.	10	9	10	11	15	8	10	14
10.	13	10	6	13	11	9	18	10

Page viii

	a	b	c	d	e	f	g	h
1.	9	5	6	10	5	8	13	3
2.	11	11	10	9	6	7	3	13
3.	8	10	12	10	9	2	18	12
4.	17	6	0	15	4	8	11	14
5.	7	10	12	7	11	11	14	1
6.	11	5	10	16	2	13	14	12
7.	7	15	8	12	10	15	9	10
8.	9	11	6	6	13	9	15	13
9.	17	8	16	8	12	10	9	7
10.	12	14	8	16	11	14	9	13

Page ix

	a	b	c	d	e	f	g	h
1.	3	4	3	1	4	6	1	6
2.	9	0	9	3	2	9	6	4
3.	5	5	5	1	1	2	2	4
4.	8	4	8	4	2	7	3	5
5.	7	3	8	1	8	8	5	6
6.	9	5	6	0	2	8	8	7
7.	4	6	8	3	0	5	6	6
8.	3	4	7	3	9	9	5	7
9.	7	2	8	2	9	5	3	2
10.	9	2	4	7	7	6	7	9

Page x

	a	b	c	d	e	f	g	h
1.	3	2	4	9	2	1	5	3
2.	2	5	9	5	2	3	4	9
3.	4	3	4	5	7	3	6	6
4.	6	4	0	9	2	3	7	6
5.	9	2	1	9	7	3	4	4
6.	5	3	4	7	4	6	7	8
7.	8	8	1	8	2	2	8	1
8.	7	6	0	8	6	1	7	8
9.	6	8	5	7	0	1	9	9
10.	9	0	8	5	6	5	2	7

Page xi

	a	b	c	d	e	f	g	h
1.	14	32	0	9	15	9	14	4
2.	48	5	10	6	0	36	63	3
3.	7	30	0	28	12	0	20	0
4.	56	16	56	18	8	25	40	27
5.	24	0	27	12	32	42	10	42
6.	21	36	24	18	48	36	49	1
7.	18	15	24	30	2	12	54	35
8.	81	20	54	0	72	12	16	8
9.	18	4	35	64	8	45	24	6
10.	40	28	16	21	6	63	45	72

Page xii

	a	b	c	d	e	f	g	h
1.	12	4	20	56	35	64	6	0
2.	18	30	9	18	9	16	49	4
3.	24	72	0	6	42	3	40	54
4.	0	12	35	24	25	56	30	12
5.	63	28	18	2	45	8	28	0
6.	15	48	32	21	20	8	0	63
7.	14	6	36	18	0	36	40	10
8.	16	15	45	0	72	12	21	8
9.	48	16	42	81	24	32	36	5
10.	27	27	0	14	10	7	24	54

Page xiii

	a	b	c	d	e	f	g
1.	4	6	2	5	6	9	1
2.	1	9	7	9	1	7	3
3.	1	3	5	0	3	6	5
4.	4	8	5	8	2	6	0
5.	9	8	9	5	0	2	4
6.	6	8	1	0	7	7	5
7.	4	3	5	3	8	1	1
8.	0	2	7	6	6	2	0
9.	4	1	5	8	7	4	7
10.	2	5	9	7	8	3	7
11.	6	4	3	2	6	3	3
12.	9	8	9	4	8	2	9

Page xiv

	a	b	c	d	e	f	g
1.	2	7	3	5	0	3	9
2.	4	1	3	6	6	4	7
3.	8	9	6	4	5	8	1
4.	5	6	1	0	6	9	2
5.	5	7	7	7	4	7	1
6.	5	5	0	8	2	2	4
7.	0	4	3	8	0	5	7
8.	2	3	6	6	8	3	3
9.	3	7	6	2	3	8	1
10.	1	4	9	8	4	2	5
11.	5	7	9	9	9	2	9
12.	8	1	4	6	2	8	9

Page 1

	a	b	c	d	e	f
1.	38	50	68	95	125	122
2.	41	48	15	85	89	179
3.	769	761	975	1390	601	1520
4.	613	433	592	527	1575	2899
5.	7788	10010	10263	17190	11011	
6.	3131	1779	44298	28693	36897	
7.	85758	59473	84125	55133	81222	
8.	51123	23008	39019	55705	69676	
9.	106	1697	9542	72937	84173	

Page 2

1. 435 ; 201 ; 636
2. 435 ; 123 ; 312
3. 759

Page 11

	a	b	c	d	e
1.	51	158	154	276	631
2.	48	12	79	169	97
3.	399	7089	3686	7845	30385
4.	14313	12812	85374	53966	80097

5. subtract ; 3926   6. add ; 62842   7. 95066

157

Page 12

	a	b	c	d	e
1.	99	192	608	3456	2468
2.	33945	2793	15504	27082	260235
3.	39483	250560	171864	2181114	4192533
4.	9	37	17 r4	325	38
5.	147 r41	135	785	2444 r8	724 r62

Page 25

	a	b	c	d	e
1.	86	342	4056	8106	50778
2.	805	3648	5115	86352	137566
3.	43776	54825	426512	1366530	1126428
4.	4	35	14 r39	116 r24	58 r20
5.	135	27 r15	1016 r32	3844 r5	434

Page 26

	a	b	c	d
1.	$\frac{3}{10}$	$\frac{16}{35}$	$\frac{10}{21}$	$\frac{4}{25}$
2.	$\frac{1}{10}$	$\frac{5}{24}$	$\frac{9}{28}$	$\frac{4}{15}$
3.	$1\frac{1}{5}$	$4\frac{4}{9}$	$4\frac{1}{2}$	$2\frac{2}{3}$
4.	$13\frac{1}{3}$	$12\frac{1}{2}$	$\frac{2}{3}$	$10\frac{1}{2}$
5.	$2\frac{11}{12}$	$2\frac{23}{56}$	$6\frac{2}{3}$	8

Page 41

	a	b	c	d		a	b	c	d
1.	$\frac{5}{12}$	$\frac{35}{48}$	$\frac{10}{21}$	$\frac{9}{64}$	4.	$13\frac{1}{3}$	$1\frac{3}{5}$	8	$3\frac{3}{4}$
2.	$\frac{10}{21}$	$\frac{28}{45}$	$\frac{3}{4}$	$\frac{1}{2}$	5.	$3\frac{17}{21}$	$6\frac{3}{10}$	$2\frac{14}{15}$	$11\frac{2}{3}$
3.	$1\frac{1}{5}$	$4\frac{2}{7}$	4	$6\frac{2}{3}$					

Page 42

	a	b	c	d		a	b	c	d
1.	$\frac{4}{7}$	$\frac{2}{3}$	$\frac{5}{8}$	$\frac{3}{5}$	4.	$3\frac{1}{4}$	$4\frac{23}{40}$	$5\frac{1}{9}$	$1\frac{3}{4}$
2.	$1\frac{1}{6}$	$1\frac{1}{4}$	$\frac{11}{24}$	$\frac{11}{40}$	5.	$2\frac{5}{18}$	$4\frac{1}{2}$	$10\frac{7}{30}$	9
3.	$2\frac{3}{5}$	$\frac{1}{8}$	$5\frac{13}{30}$	$3\frac{9}{10}$					

Page 53

	a	b	c	d		a	b	c	d
1.	$\frac{7}{8}$	$\frac{1}{2}$	$1\frac{1}{6}$	$\frac{2}{9}$	4.	$5\frac{13}{30}$	$3\frac{11}{40}$	$5\frac{13}{24}$	$6\frac{2}{3}$
2.	$1\frac{1}{18}$	$1\frac{5}{24}$	$\frac{1}{2}$	$\frac{1}{12}$	5.	$3\frac{17}{20}$	$3\frac{1}{8}$	$12\frac{59}{120}$	$11\frac{7}{24}$
3.	$5\frac{8}{9}$	$2\frac{3}{7}$	$5\frac{5}{24}$	$3\frac{5}{8}$					

Page 54

	a	b	c	d
1.	8	$10\frac{1}{2}$	10	$10\frac{1}{2}$
2.	$\frac{1}{8}$	$\frac{3}{10}$	$\frac{1}{7}$	$\frac{4}{27}$
3.	$\frac{2}{7}$	$\frac{1}{2}$	2	$1\frac{1}{6}$
4.	$1\frac{1}{4}$	$\frac{9}{10}$	2	$1\frac{1}{3}$
5.	$\frac{3}{4}$	$2\frac{1}{2}$	$\frac{3}{14}$	$2\frac{1}{7}$

Page 65

	a	b	c	d		a	b	c	d
1.	15	$10\frac{2}{3}$	6	$11\frac{2}{3}$	4.	$1\frac{1}{2}$	$1\frac{1}{15}$	$1\frac{1}{6}$	$\frac{3}{4}$
2.	$\frac{1}{6}$	$\frac{4}{21}$	$\frac{1}{9}$	$\frac{3}{28}$	5.	$\frac{7}{15}$	$3\frac{3}{5}$	$\frac{2}{9}$	$2\frac{2}{9}$
3.	$\frac{4}{9}$	$\frac{1}{2}$	2	$1\frac{1}{3}$					

Page 66

	a	b	c
1.	.7	3.19	5.025
2.	.8	3.32	3.128
3.	$\frac{4}{5}$	$9\frac{33}{100}$	$16\frac{1}{8}$

	a	b	c	d
4.	.8	1.07	8.023	14.076
5.	1.1	.76	.438	10.274
6.	1.08	1.086	7.525	46.208
7.	.32	3.317	3.55	.285

Page 81

	a	b	c
1.	.175	9.4	3.08
2.	.90	3.2	5.300
3.	$\frac{3}{40}$	$8\frac{3}{5}$	$16\frac{49}{100}$

	a	b	c	d
4.	1.3	.95	14.361	30.646
5.	1.7	4.51	.067	5.697
6.	1.32	.415	18.273	57.355
7.	.166	2.59	2.784	4.825

Page 82

	a	b	c	d	e
1.	3.5	.8	3.6	1.2	3
2.	.42	.06	.32	.72	.4
3.	.012	.008	.006	.004	.012
4.	.21	.08	.24	.54	.4
5.	.048	.005	.006	.0042	.0004
6.	4	.8	67.2	234	5680
7.	4.8	.235	.272	1.2525	.585

Page 91

	a	b	c	d	e
1.	.48	1.26	.2772	2.226	4.844
2.	5.4	.72	29.82	.48	75.92
3.	.015	.576	2.835	35.38	14.952
4.	.08	.212	.2408	118.712	162.8
5.	.0054	.0134	37.24	15.5944	264.176

Page 92

	a	b	c	d
1.	7.3	.27	.045	.0019
2.	20	150	2100	4000
3.	1.2	21	2.4	5
4.	80	60	3630	600
5.	150	.14	.46	.92

Page 105

	a	b	c	d
1.	.023	.14	3.7	.0051
2.	1100	12000	60	2100
3.	23	1.4	2.4	7
4.	900	70	90	10300
5.	25	1.4	2.6	.073

Page 106

	a	b		a	b
1.	7000	.085	5.	3500	.0065
2.	190	.045	6.	8200	.255
3.	6000	.007	7.	289	8.5
4.	5000	4	8.	90	45

Page 117

	a	b		a	b
1.	9000	.365	6.	900	.785
2.	720	.084	7.	2000	998
3.	26000	.017	8.	1.5	450
4.	4510	.065	9.	25.2	9.8
5.	6800	3.5	10.	240	70.2

Page 118

	a	b		
1.	36	4	9.	56
2.	18	8	10.	15
3.	10560	5280	11.	38
4.	6	7	12.	150
5.	14	2		
6.	32	1		a	b
7.	360	3	13.	28	18
8.	2	2	14.	330	

Page 127

	a	b		a	b
1.	96	6	9.	70	48
2.	15	10	10.	17	3
3.	10560	1760			
4.	12	$5\frac{1}{2}$	11.	76
5.	36	6	12.	210
6.	20	2	13.	4320
7.	96	2	14.	a	b
8.	300	2		56	45

Page 128

	a	b
1.	7	90
2.	35	52
3.	7	40
4.	13.5	135
5.	.06	.67
6.	.0625	1.25
7.	$\frac{9}{100}$	$\frac{1}{5}$
8.	$\frac{9}{20}$	$\frac{14}{25}$
9.	.59 or $\frac{59}{100}$	7.5 or $7\frac{1}{2}$
10.	84	40.05 or $40\frac{1}{20}$
11.	36	122.01 or $122\frac{1}{100}$
12.	21.6 or $21\frac{3}{5}$	63
13.	33.75	8.194
14.	52.5	5.551

Page 137

1.	.03	3%	11.	16
2.	.25	25%	12.	62.4 or $62\frac{2}{5}$
3.	.35	35%	13.	43.75
4.	$\frac{3}{50}$	6%	14.	20.28
5.	$\frac{39}{100}$	39%	15.	8.228
6.	$\frac{1}{8}$	12.5%
7.	$\frac{1}{20}$	.05
8.	$\frac{7}{25}$	.28
9.	$\frac{3}{4}$	.75
10.	$\frac{9}{10}$	.9

Page 138

1. c 3. h 5. g
2. d 4. a 6. f

Page 142

1. f   4. h   7. a
2. e   5. g
3. d   6. b

Page 143

	a	b	c	d	e
1.	1067	181170	1214	19061	10837
2.	1736	8088	17550	6355956	6633583
3.	12	2971 r2	234 r16	4163 r15	207 r22
4.	12.4	1.486	$45.85	$3.29	8.612
5.	$\frac{5}{24}$	$\frac{2}{7}$	$5\frac{1}{3}$	7	4
6.	1	$1\frac{1}{6}$	$1\frac{17}{24}$	$5\frac{7}{12}$	$12\frac{11}{12}$

Page 144

7.	$2\frac{1}{2}$	2.5	2.50	2.500
8.	$4\frac{1}{10}$	4.1	4.10	4.100

	a	b	c	d	e
9.	$\frac{2}{7}$	$\frac{3}{10}$	$7\frac{1}{8}$	$2\frac{1}{4}$	$1\frac{17}{24}$
10.	24	$\frac{2}{35}$	$1\frac{1}{6}$	$\frac{1}{2}$	$1\frac{1}{2}$
11.	$\frac{1}{2}$	12. $1\frac{3}{4}$
13.	$113.76	14. 3.264

Page 145

	a	b	c	d	e
1.	7422	9764	93444	119	5431
2.	1678	71273	1602	10146	236655
3.	499410	49 r1	1720 r5	365	1109 r32
4.	9.94	28.287	1.0052	37.1	.27
5.	.375	6.056	.129	157.95	9.295
6.	.009	4.15	4.25	2.4	90

	a	b	c	d
7.	$\frac{1}{5}$	$4\frac{1}{5}$	22	20

Page 146

	a	b	c	d
8.	$1\frac{1}{4}$	$1\frac{1}{2}$	$7\frac{3}{12}$	$12\frac{7}{12}$
9.	$\frac{3}{5}$	$\frac{3}{8}$	$4\frac{1}{6}$	$1\frac{7}{24}$
10.	36	$\frac{7}{48}$	$\frac{10}{21}$	$2\frac{1}{4}$

| | a | b | | a | b |
| 11. | 135 | $102\frac{1}{2}$ | 12. | 52.9 | 1296 |

Page 147

13.	$\frac{1}{10}$; .1	18. d
14.	75% ; .75	19. g
15.	50% ; $\frac{1}{2}$	20. b

	a	b	21. e
16.	6	14.28	22. f
17.	25.6	380

Page 148

	a	b		
23.	230	2350	28.	4.66
24.	2000	.678	29.	30560
25.	2	112	30.	$1\frac{7}{8}$
26.	3	16	31.	1483
27.	80	315	32.	$37\frac{3}{4}$

159